DATE DUE			

Introduction to Safety in the Chemical Laboratory

Introduction to Safety in the Chemical Laboratory

by

N. T. Freeman
Deputy Chairman
Efficiency Aids Ltd
Stockton-on-Tees
Cleveland, UK

J. Whitehead
Chief Analyst
Tioxide International Ltd
Stockton-on-Tees
Cleveland, UK

ACADEMIC PRESS, INC.
(Harcourt Brace Jovanovich, Publishers)
London Orlando San Diego New York
Toronto Montreal Sydney Tokyo

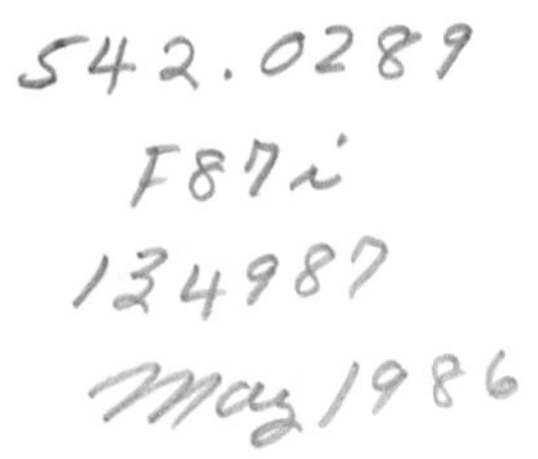

ACADEMIC PRESS INC. (LONDON) LTD.
24/28 Oval Road
London NW1

United States Edition published by
ACADEMIC PRESS, INC.
Orlando, Florida 32887

British Library Cataloguing in Publication Data

Freeman, N. T.

Introduction to safety in the chemical laboratory.
1. Chemical laboratories—Safety measures
I. Title II. Whitehead, J.
363.1'79 QD51

ISBN 0-12-267220-8

LCCCN 82-45029

PRINTED IN THE UNITED STATES OF AMERICA

84 85 86 87 9 8 7 6 5 4 3

Preface

In the first twenty or thirty years of this century laboratories were normally found only in the chemical industry, in hospitals, or in educational establishments; now they are commonplace. Few industrial locations are without laboratory facilities of a size compatible with the range of operations and the substances being handled or used. Rapid changes and advances in the scientific fields of research and development call for ever larger, more modern, and better equipped laboratories if enterprises are to keep pace with, or outstrip, competitors. Moral and legislative requirements now impose on the manufacturers and suppliers of products heavy responsibilities with regard to consumer protection. The health and safety of a company's own employees have always been important considerations, but the users, manufacturers, shippers, and importers of chemicals must now not only safeguard those in their own work environment: their duty extends to the employees of customers and to the public in general. This extension of responsibility has been accompanied by an increasing concern with, and awareness of, the quality of life in general.

Industrial chemists and technicians, through their work in laboratories, provide, and will continue to provide, a platform from which the discharge of these increasingly onerous responsibilities can be facilitated. As the scope, size, and equipment of industrial laboratories become larger and more sophisticated so too must that in educational establishments in order that the skills of emergent laboratory workers keep pace with progress. Hospital, pathological, and forensic laboratories now operate on a scale unthought of a couple of generations ago.

This upsurge in activities has been accompanied by useful literature intended to raise levels of health and safety. Generally this has taken the form of helpful and voluminous lists of hazardous substances, threshold limit values, compatible and incompatible chemicals and so on. We have dealt with the general approach to safety—how to design accident prevention into the laboratory; coping with hazards which arise almost every working day; dealing with actual or potential fires and disasters, and similar matters. It was an apparent lack of collected information of this kind which inspired the authors to write this book, which should be taken as a guide rather than a manual of procedures. The aim is to provide a book of reference for everyone who works in, or uses the facilities of, laboratories.

ACKNOWLEDGEMENT

The authors wish to acknowledge with thanks the help received from: Audrey Bodley, Andrew Gibson, and Louise Gibson.

DISCLAIMER

The authors of this book have each been involved in laboratory safety activities for more than thirty years. The sources used to supplement their own practical experience are believed to be reliable and to represent the best opinion as at 1981. The book is intended as a guide rather than a manual of procedure. No warranty, guarantee, or representation is made by the authors and publishers as to the correctness or adequacy of any information contained therein. No responsibility or liability is accepted in any manner whatsoever for any errors or omissions.

Contents

1

Laboratory Design and Layout

It is beyond the scope of this book to deal with special-purpose laboratories such as those designed for radio-chemical work, animal studies,or toxicological investigations, and the references at the end of this chapter should be consulted for detailed information of this type. It is intended that this chapter will include discussion on basic principles of laboratory design with particular emphasis on safety. A laboratory is defined as a building set aside for experiments in natural science and, as such, may involve work in many scientific disciplines of which chemistry, physics, mathematics and engineering are examples. The general-purpose chemical laboratory contains a wide range of potentially hazardous situations. Flammable solvents introduce the risks of explosion and fire; toxic chemicals the risk of death by poisoning which can be almost instantaneous or take many years to happen; corrosive chemicals can cause serious injury and disfigurement, and extremely lethal sources of electricity such as those used to power X-ray tubes can bring about electrocution. Considerable progress has been made in the last few years in the development of new materials for use in laboratory construction and there is now a very wide choice of finishes for walls, floors, and benches. In selecting materials, particular attention must be directed towards eliminating hazards and ensuring the safety of personnel working in the laboratory.

Overcrowding is one of the significant causes of accidents in laboratories. Adequate space must be allowed for each worker but clearly the exact amount will depend on the type of work done. The British Standard Specification entitled Recommendations on Laboratory Furniture Fittings (BS 3202 1959) gives some guidance, suggesting that a space of under 9 m^2 per worker results in inefficient cramping whereas over 23 m^2 is seldom required. In the case of school laboratories, between 2.75 and 4.5 m^2 per pupil is considered adequate. For an analyst carrying out classical methods of analyses a bench length of 3 m is adequate. An examination by CSIRO (Australia),[1] however, indicated that in eight of their laboratories the average space available ranged between 41 and 50 m^2 and in another four, between 57 and 61 m^2. They conclude that a figure of 50 m^2 should be used but added a rider justifying a greater area for some types of work. In the case of ventilation, BS 3202 1959 recommends that at least 57 m^3 of fresh air per person per hour should be provided.

TYPE OF BUILDING

Although this chapter is devoted primarily to the internal construction and design of laboratories, some details on the important points in the construction of the actual building are included. Clearly, during the design stage consultations will take place with the relevant statutory bodies concerned with building construction and fire hazards to ensure compliance with the appropriate regulations. Discussions should also take place with the insurers of the property as they may enforce additional special conditions before they will provide cover. (Construction details in relation to fire protection are dealt with in Chapter 9.)

In defining areas of special hazards, such as in radiation laboratories, access to the roof space above the laboratory must be restricted. Laboratories in high-rise buildings are usually placed on the upper floors in order

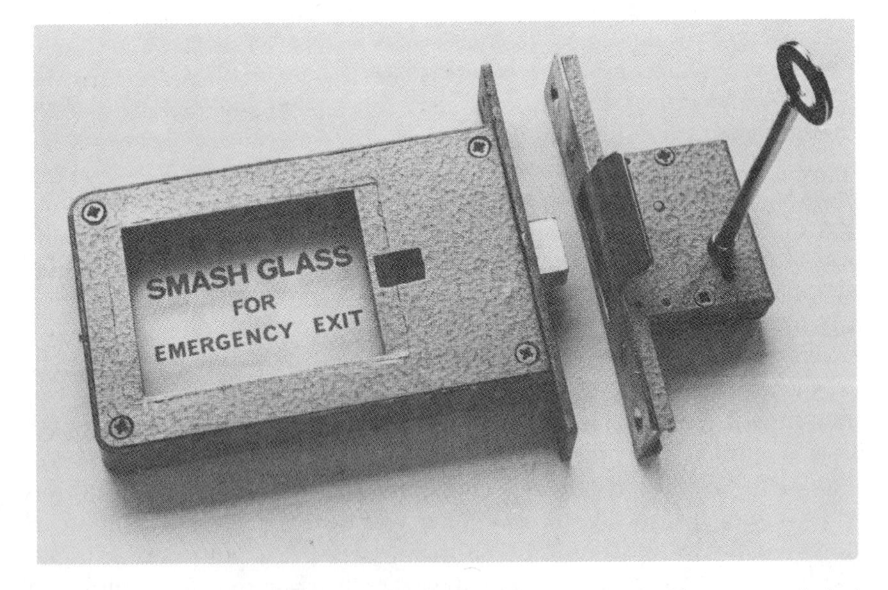

Fig. 1.1. A lock that provides security and enables occupants of the building to open the lock without a key by smashing the glass plate. It allows strong springs to withdraw the bolt thus enabling the door to be opened. The door cannot be relocked until another glass panel has been inserted. Routine opening of the door is possible using the key locking strike. (Photograph courtesy of Albert Marston & Co. Ltd).

to reduce the length of fume extraction ducting and the risk of fumes entering the building; in the disposal of liquid effluents the containment of liquid is simpler and the risk of contamination less. There are, however, problems in transporting corrosive and toxic materials to an upper floor, and the latest UK legislation on the housing of compressed gas cylinders may lead to a reconsideration of the siting of laboratories in multi-storey buildings. In view of their exposed position, additional precautions concerning such fire-fighting equipment as compressed air breathing equipment, are desirable, and at least one of any lifts installed should be large enough to take a casualty on a stretcher. Emergency exits must be available for use at any time and not be permanently locked. Various devices are commercially available for securing such exits (Fig. 1.1) and of these the easily-operated pushbar type is preferred to a locked door where the key is housed nearby in a glass-fronted case.

Other points concern the siting of stores to facilitate access and distribution, and the width of corridors, which need not exceed 2 m even in the largest laboratory block as greater width encourages storage in them. Detailed points regarding materials of construction, furniture, fume cupboards, lighting, and services will be included in later sections.

STRUCTURE

Walls should have a slow flame-spread characteristic, be smooth and readily cleaned, and not have any ledges that would form dust traps. A plastered finish brick or breeze block wall painted with an alkyd resin paint is ideal, although if extreme toughness of finish is required a polyurethane paint may be used. For general use, PVC floor covering is a long-lasting durable material which has a high resistance to acids and alkalis. It is, however, attacked by solvents such as acetone and chloroform and is unsuitable if large amounts of these solvents are in use. In this case, ceramic tiles would be a satisfactory, if more expensive, alternative. There is an increasing tendency, particularly in laboratories with expensive instruments such as electron microscopes and X-ray equipment, to use synthetic fibre carpeting. This can be vacuum-cleaned, thereby reducing dust which can give trouble if it gets into delicate instruments. Whatever the material, the undersurface should preferably be of a rigid finish such as concrete, and the floor covering should be seamless with coves formed at the junction with the wall to help contain spillages. There is some merit in floor tiles instead of sheeting since worn patches can be easily and economically replaced. In laboratories where heavy equipment is being handled a stronger floor finish is desirable and a concrete floor treated with one of the proprietary two-pack epoxy coating materials for sealing concrete is advantageous. Floor loading tests are needed for laboratories above ground level. Suspended tile ceilings using materials of acceptable flame resistance are attractive in appearance and aesthetically pleasing. Services can be hidden behind them, but the possibility of undetected incipient fire is then created. Alternatively, a traditional painted plaster ceiling can be used. Doors to laboratories should be fitted with observation panels of a type selected with regard to the fire and impact risks in the area.

BENCHES

Benches are of three types: wall, peninsula, and island. Most laboratories have a combination of wall and one of the other types of bench. The type of bench selected depends on the work being carried out in the laboratory, the available space, and the financial budget. Where work of a high hazard rating is being performed, island benches are to be preferred because they avoid the cul-de-sac areas created by peninsula benches. However, the latter reduce the number of people passing by, which can be an advantage when delicate work is being performed. The dimensions generally used for benches are: height, 0.91 m; width, 0.75 m for wall type and 1.5 m for island or peninsula types. Materials used for the tops of the benches include teak or a similar hardwood such as iroko, afrormosia, oak, formica, ceramic tiles, stainless steel, and pyroceram. Selection of the type of material to be used depends on the type of work being carried out in the laboratory. For general purposes, the ceramic tile finish is excellent and details of a typical bench are shown in Fig. 1.2. Particular attention is drawn to the lip on the edge of the bench which retains any spilled liquids. The services to the bench are preferably led along a central spline which is easily seen in Fig. 1.2. The traditional high reagent shelf has been dispensed with; this eliminates the risk of bottles falling off the top shelf and also increases the feeling of spaciousness in the laboratory. The type of bench illustrated is supported on a rigid wooden base with either drawer or cupboard fitments. The drawers should be fitted with stops to prevent them being withdrawn completely. In cases where instrumentation is used extensively, it is difficult to service the instruments if access to their rear is necessary. In this case a special split bench has been designed by the laboratory staff (at Tioxide International Ltd) to facilitate all-round access to the instruments as shown in Fig. 1.3. The illustration shows a bench which has been specially designed for atomic absorption spectrometry. The service outlets, which comprise cold water, electricity, compressed air, vacuum, acetylene, nitrous oxide, nitrogen, and argon are situated at each side of the centre aisle. Acetylene is led to the bench from the cylinders external to the building via the exposed downpipe—four outlets with flash arrester units can be seen on the top of the benches. The distinctively shaped handwheel for the acetylene valves can also be seen. Special localized fume extraction equipment is placed above the bench to remove the heat and fume from the flames. The rack on the wall holds the hollow cathode lamps which are used in this method of analysis. Sinks are usually placed at the ends of benches, tun dishes being used to receive water from taps situated along the length of the bench. These are preferred to central troughs running along the bench which are difficult to keep clean. Choice of material for the sink is dependent on the chemicals most in use in the laboratory. If substantial amounts of hydrofluoric acid are used then stainless steel is preferable to porcelain. Draining boards should be placed alongside the sinks; a typical sink installation at the end of a double bench is shown in Fig. 1.2. The draintrap is placed below the sink

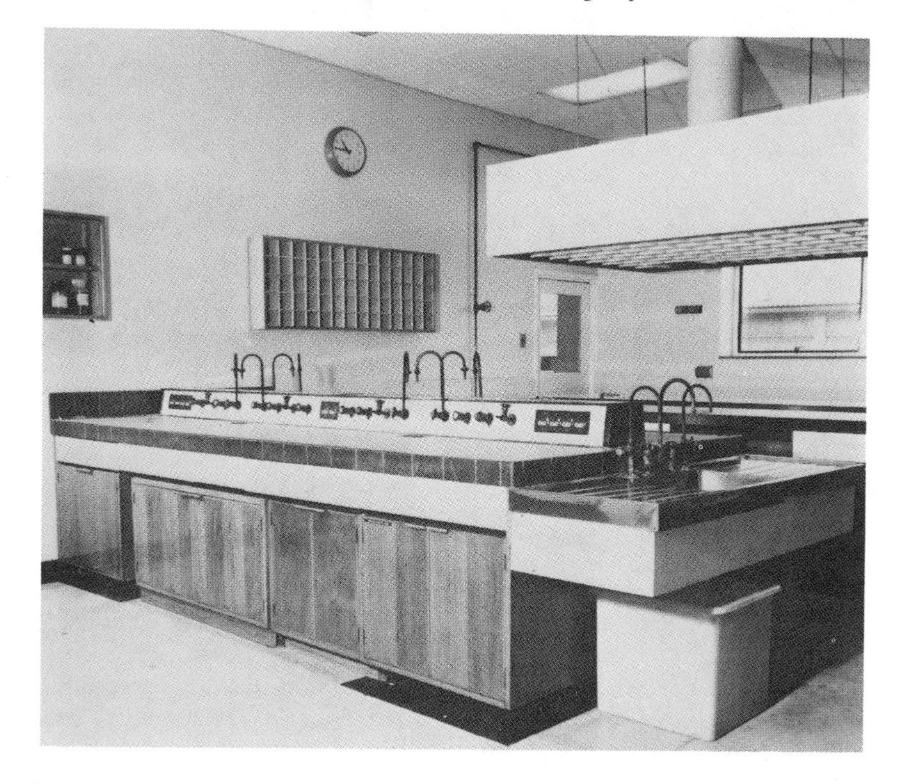

Fig. 1.2. A typical modern laboratory bench. Services are led from the wall side along the central spline. The sink unit is stainless steel and the drain trap is easily accessible at the end of the bench. Ceramic tiles are used for the bench top and cupboards or drawers can be placed underneath. A rubbish bin is housed under the sink.

where it is readily accessible, and the space below the sink is used to house waste-bins.

Fig. 1.3. A split bench designed for work with instruments where access to the rear is important. Here the benches are intended to provide space for 4 atomic absorption spectrometers and the special gas supply arrangements can be seen. These include acetylene for which the specially distinguishable handwheels and flame arresters can be seen. The acetylene supply which UK regulations specify must not be concealed is visible on the wall. The arrangement above the bench conceals a fume exhaust system and the wall rack houses the hollow cathode lamps.

STORAGE

Storage space is an item which frequently gets overlooked in laboratory design. It is exceptional to find that all the equipment is in use at any one time and there is usually a need to store fairly large amounts of equipment and chemicals, as well as materials that have been produced or samples that have been analysed. It is worth, therefore devoting some time to planning storage space as untidy stacking of equipment and materials in a laboratory is a frequent cause of accidents. Note that, at this point, the primary concern is storage of materials in the laboratory and external stores are not being considered here.

With regard to chemicals, the amount stored in a laboratory must be kept to a minimum. Highly flammable materials defined in the UK as those with a flashpoint of less than 32°C should be kept in closed containers (preferably metal if the quantity is greater than one litre) and stored when not in use in metal cabinets of approved design. Although UK regulations permit up to 50 litres to be stored in the workroom, efforts should be made to keep the volume well below this. Corrosive chemicals should be stored as near to the floor level as possible and in non-corrodible trays. Hand bottles containing corrosive chemicals should also be kept in trays. Toxic chemicals should be stored in well-ventilated areas and preferably in a fume cupboard or hood. Poisonous chemicals should be kept in a locked cupboard and under rigid key control. Chemicals safeguarded in this way should be those included in a UK Schedule[2] or, in general terms, any chemical having the "poison" symbol on label. Certain chemicals, e.g. powerful oxidizing agents and organic compounds, react violently together and these must be stored apart. Sodium hydroxide and yellow phosphorus in stick form are very much alike and should not be stored alongside each other.

Hazards in the storage of equipment are not so prevalent, but thought given to reserving cupboards for beakers, volumetric flasks, funnels, etc., will pay dividends in the saving of time. The drilling of holes in cupboard shelves greatly facilitates the storage of separating funnels, and specially lined and fitted drawers are recommended to accommodate fragile glass equipment such as burettes and pipettes.

Finally, the need to dispose of rubbish is frequently overlooked. The problem of designing rubbish containers into the furniture of the laboratory so that they do not look obtrusive and yet are readily accessible is not easy. One way of doing this is illustrated in Fig. 1.2 in which the bin is placed under the sink at the end of the bench; another solution is to put the container in place of a cupboard under the bench. A special bin clearly marked should be reserved for disposing of broken glass. Corridors must not be used for storing equipment other than fire extinguishers, first-aid materials, and breathing apparatus which should be placed in clearly marked cabinets, preferably inset into the walls.

SERVICES

In the event of a hazardous situation developing in the laboratory it is essential that all the services are able to be isolated at a central point which is preferably outside the door. Such an arrangement is shown in Fig. 1.4. This also facilitates maintenance of services within the laboratory. The services to the modern laboratory are numerous and can consist of: water, steam, compressed air, natural gas, acetylene, argon, argon/methane, butane, carbon dioxide, helium, hydrogen, nitrogen, nitrous oxide, oxygen, and vacuum. In fact in the authors' laboratories there are seven different types

Table 1.1a. Colour identification system for pipelines.

Pipe Contents	*Basic Identification Colour*	*BS Colour Reference BS 4800 1972*
Water	Green	12D 45
Steam	Silver-grey	10A 03
Mineral, vegetable and animal oil; combustible liquids	Brown	06C 39
Gases in either liquid or gaseous condition (except air)	Yellow ochre	08C 35
Acid and Alkalis	Violet	22C 37
Air	Light blue	20E 51
Other fluids	Black	Black
Electrical Services	Orange	06E 51

Table 1.1b. Safety colours.

Safety Colour	*BS Colour Reference BS 4800 1972*
Red	04E 53
Yellow	08E 51
Auxiliary blue	18E 53

The basic identification colour is used to identify the material in the pipe and the safety colour indicates the type of hazard, e.g. a pipe containing (a) cold drinking water would be painted green with band coloured auxiliary blue; (b) fire extinguishing water would be painted green with a red band; (c) a gas containing a radioactive compound would be painted yellow ochre, with a yellow band on which is placed the radioactive symbol. (Material from BS 1710, 1975 is reproduced by permission of the British Standards Institution, 2 Park Street, London W1A 2BS, from whom complete copies can be obtained.)

Fig. 1.4. A typical services isolation unit shown with the door open and closed. The colour code for the pipelines is attached to the front of the door.

of water supply: hot, process cold, condensate, demineralized, fire hydrant, douche, and drinking cold. The pipelines can be colour-coded using one of the nationally recognized systems (see Table 1.1): Identification of Pipelines BS 1710 1975 or Identification colours for pipes conveying fluids in liquid or gaseous condition in land installations and on board ships ISO R508 Table I, 1966. However, it is also a wise precaution to name the pipes as well, as colours can fade; a colour code system can also be used on plastic inserts on the handle of each control valve on the bench and a system for doing this is given in the German DIN standard No. 12920, 1971 (see Table 1.2). In the case of particularly hazardous gases such as acetylene, a specially shaped handwheel is advantageous. Where special compressed gases are in frequent use in a laboratory it is preferable to pipe these on to the bench from gas cylinders which can then be housed in a safe area outside

Fig. 1.5. A liquid nitrogen reservoir; as well as being a source of liquid from the tap at the front, it also supplies gaseous nitrogen through the tube system at the rear of the tank as a piped service to a large laboratory block.

Table 1.2. Colour code for fluids on hand levers and handwheels of laboratory taps – DIN12920, 1971.*

Materials	*Colour of Zone No.*		
	1	*2*	*3*
Towns Gas	Zones and handle all yellow		
Propane/Butane	Yellow	Red	Red
Hydrogen	Yellow	Red	Red
Nitrogen	Yellow	Green	Green
Carbon Dioxide	Yellow	Grey	Grey
Argon	Yellow	Grey	Yellow
Helium	Yellow	Grey	White
Methane	Yellow	Red	Blue
Acetylene	Yellow	Red	White
Ethylene	Yellow	Red	Black
Propylene	Yellow	Red	Green
Compressed Air from compressor	Blue	Grey	Grey
Oxygen	Blue	Blue	Yellow
Vacuum, high	Grey	Grey	Grey
Steam <2.5 bar	Red	Red	Red
Steam >2.5 bar	Red	White	Red
Water, cold drinking	Green	Green	Blue
Water, hot drinking	Green	Green	Red
Water, deionized	Green	White	Blue
Water, distilled	Green	Red	Blue

*Material reproduced by permission of Deutshes Institut für Normurg e.v. Postfoch 1107, D-1100 Berlin 30, from whom complete copies can be obtained.

the laboratory. Bulk supplies of gases such as oxygen, nitrogen, argon, and liquefied petroleum gas can be obtained and it should be noted that tanks of liquid nitrogen, for example, can be used to supply piped gaseous nitrogen as well as the liquid. Such an installation is shown in Fig. 1.5. Liquid nitrogen can be withdrawn from a valve at the front of the tank while the tube arrangement at the rear gasifies the nitrogen for supply to the laboratories. The placing of the service outlets along the bench will depend on the type of work in progress; however, it is always wise to err on the generous side, particularly with electrical outlets. In the case of open benches, the controls for all the services may be mounted on the central service panel. Special arrangements are necessary for fume hoods or cupboards and these are described on page 14.

Combustible gas supplies which are used in conjunction with compressed air or oxygen must be fitted with a non-return valve to avoid explosions occurring because of air or oxygen being forced into the gas line.

It is a fairly simple matter to bring the services to a peninsula bench either along the wall, down from the ceiling, or upwards from the floor. In all these cases, the pipework can be concealed from view quite successfully. The services to an island bench are either led along a duct in the floor (which can lead to problems with spillages) or from the ceiling which, unless the pipes are specially housed, is unsightly. Note that in the case of acetylene the pipework must always be in view and must never be of copper or a copper alloy.

Waste outlets are now invariably made from polythene and it is good practice to incorporate a glass trap so that blockages can readily be seen.

LIGHTING

Fluorescent fittings are now almost universally used for lighting in laboratories. For general laboratory work, a minimum intensity of 538 lumens/square metre (50-foot candles) is necessary, and an intensity of twice this level is preferable. If localized higher intensity lighting is required then this is best provided by the use of spotlights but care must be taken that these do not provide a hazard due to trailing cables etc. Fifty-four to 107 lumens/square metre intensity is adequate for corridors. In hazardous areas flame-proof fittings are required and it may also be desirable to install emergency lighting. Inexpensive units are now available in which a battery-powered light is switched on automatically in the event of a mains supply failure.

VENTILATION AND FUME EXTRACTION

Good ventilation to prevent the build-up of dangerous concentrations of flammable or toxic gases is essential in a laboratory. Fairly straightforward systems are employed in laboratories which do not have fume cupboards or hoods, but where these are installed the systems become more complicated. In the case of large fume cupboard installations, when they are switched on they abstract a large volume of air from the laboratory and arrangements have to be made to replace this with a similar volume of temperature conditioned filtered air. There is a design of fume cupboard which avoids drawing air from the laboratory but this still needs filtered air to avoid contamination of articles in the cupboard. Ventilation of a laboratory must therefore always be considered together with the design of the fume extraction systems. Forced draught ventilation systems using filtered and temperature conditioned air are undoubtedly the most satisfactory for use in laboratories. The inlet air can be purified by passage through a water curtain and electrostatic precipitators and, if necessary, further filtration can be carried out at the inlet to the laboratory. The rate of air change in a laboratory will be dependent on the rate of evolution of toxic or flammable vapours but in those cases where these are low, four to six air changes per minute is the acceptable rate. The final outlet stack should be correctly sited and high enough to allow adequate fume dispersal.

Fume extraction systems

There are two general-purpose systems in use—the cupboard and the hood. The former consists of a three-sided compartment with a sliding front and the work takes place inside the compartment which can be virtually sealed off by pulling down the front. The hood is similar to a cupboard but without the sliding front and possibly one or two of the sides. Note that this terminology applies in the UK; in the USA, the term "hood" is also applied to a fume cupboard. The hood is used for pulling air away from a fairly large general area and as such is very susceptible to draughts in the laboratory. As fume cupboards may be designed to provide an equally large unencumbered area for work it is recommended that they be installed instead of fume hoods. A great deal of effort has been put into the design of fume cupboards, as evidenced by the extensive literature on the subject. The type of fume cupboard and exhaust system selected for installation will depend on the nature of the work for which it is required and before discussing details a few general principles will be given.

The position of the fume cupboard in a laboratory is important. By definition, the cupboard will be used for hazardous operations and so must not be positioned near doors as draughts from these and the passage of individuals will upset the air flow into the cupboard. Separate exhaust systems are required for each unit to avoid the possibility of back diffusion from one cupboard to another. The cupboard must be made from materials which are non-absorbent and non-combustible. Any internal lighting must be flameproofed and resistant to corrosion; the possibility of auto-ignition of highly flammable solvents such as carbon bisulphide vapour on hot surfaces must not be overlooked and for this reason fluorescent fittings are preferred. The possibility of dirt falling down the exhaust duct should be guarded against by installing a baffle inside the cupboard. The ducts and exhaust fan should be readily accessible for cleaning. The exhaust ducts should preferably go vertically to the atmosphere; bends and horizontal runs which can act as dirt traps should be avoided where possible. The outlet of the exhaust duct should go to atmosphere well away from any inlet duct to the laboratory. Exhaust fans should be flameproof and resistant to corrosion. Cupboards should not be used for experiments in which large volumes of toxic or corrosion gases are released. In these cases, it is preferable to use a dedicated extraction system together with absorption of the hazardous component.

The controls for the services should be placed outside the fume cupboard to eliminate the need for entry. In the case of electrical sockets, a compromise must be reached, for if they are placed inside (with the switches outside) they are liable to corrosion, whereas if they are outside the trailing leads may present a hazard if they are caught by passing persons.

The front and sides are usually glazed so that operations can be watched when the sash is down. Toughened, wired, and laminated glass as well as plastics have all been suggested and have their advantages and disadvantages. Plastic materials may be discounted because of their flammability and thermoplastic characteristics; toughened glass, although very strong, can easily shatter if the toughened film is penetrated; laminated and wired glass may be dangerous in violent explosions. Glass in any case is etched if considerable amounts of hydrofluoric acid are used. Attempts to prevent this in the authors' laboratories by using plastic-coated glass were unsuccessful due to discoloration of the plastic films. On balance, laminated or wired glass is recommended and if hydrofluoric acid causes troublesome etching then the only solution is to replace the glass at regular intervals. Ordinary household window glass must not be used.

Many materials have been used in the construction of the frame and the actual choice will be dependent on cost limitations and on the type of work. Hardwood is probably the cheapest material and it can be treated with a flameproof agent if necessary. Apart from PVC, plastics are not recommended because of their poor temperature resistance. Polyurethane treated asbestolite, PVC, and stainless steel are frequently used for the top and side surfaces; any of the materials described under the section on benches can be used for the working surface.

Special precautions are necessary in selecting materials of construction when reactions involving perchloric acid are carried out in fume cupboards because vapours of this acid cause wood and other organic materials to become highly flammable and many metal perchlorates are highly explosive. Precautions have to be taken to ensure that no wood or unprotected metal surfaces in the fume cupboard or in the whole extraction system are exposed to perchloric acid vapour. Brick and glass are suitable materials for the cupboard, and the ducting and fan can be made from PVC. Care should be taken to ensure that there are no crevices on which dust can accumulate and the whole system should be frequently washed down with water. An

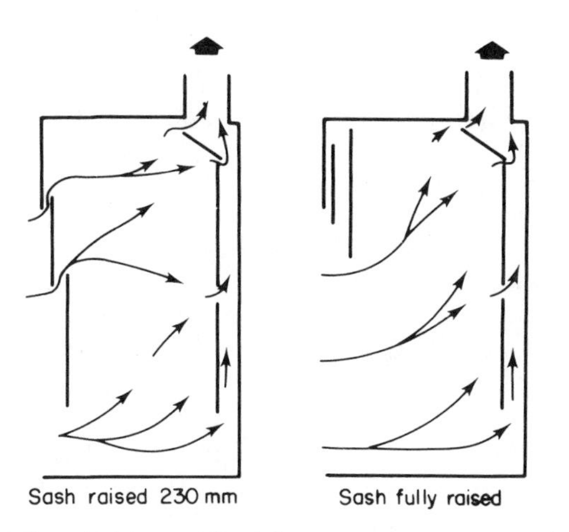

Fig. 1.6a. A fume cupboard with a stabilized flow extraction system. As the sash is lowered, a bypass is opened at the top of the cupboard to admit air from the laboratory, thereby maintaining the flow across the face at a roughly constant value. (Photograph courtesy of A. Gallenkamp & Co. Ltd).

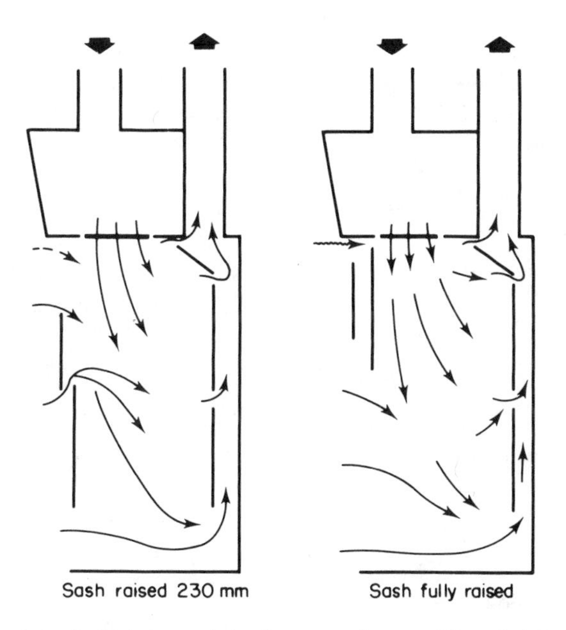

Fig. 1.6b. A cupboard which, instead of using expensive conditioned air from the laboratory to compensate, as the sash is closed a secondary source of filtered air is supplied through its roof. (Photograph courtesy of A. Gallenkamp & Co. Ltd).

alternative, if the amount being used is small, is to use a localized glass extraction system and to scrub the gases in sodium hydroxide solution.

Air Velocities

Various values are quoted for the lineal air velocity through the fully open front of a fume cupboard. Those quoted in Recommendations on Laboratory Furniture, BS 3202 1959, which are shown in Table 1.3, are typical of the values given in the literature. When the sash is closed, however, the lineal air velocity increases and can become troublesome. Systems have been introduced to circumvent this by reducing the air flow across the face of the cupboard as the sash is closed and replacing this with air from the laboratory (Fig. 1.6, a) or from outside the laboratory (Fig. 1.6, b). In the former case it is frequently necessary to introduce into the laboratory an auxiliary supply of conditioned air to compensate for the abstraction by the fume cupboard, which tends to be wasteful. The case illustrated in Fig. 1.6, ii avoids this and if the air that is introduced into the fume cupboard is carefully filtered there should be no problems from contamination. The baffle plate at the rear of the fume cupboard materially reduced the degree of turbulence. The efficiency of extraction systems for fume cupboards should be tested annually by measuring the gas flows with an anemometer. Their effectiveness in dealing with smoke should also be tested using a smoke generator with the sash in various positions from fully closed to fully open and an operator standing in the normal position of work in front of the fume cupboard. The latter point is important because the body will have an effect on the air flow pattern and may even cause eddies of fume to issue from the fume cupboard.

Many reputable companies are in the business of manufacturing fume cupboards and they should be consulted when a new laboratory is being planned. Fig. 1.7 shows a suite of fume cupboards which illustrate most of

Table 1.3. Linear air velocities through the fully open fronts of fume cupboards – BS 3202, 1959.*

Type A	Fume cupboards in schools	6–12 m min^{-1}
Type B	Fume cupboards in academic and general industrial laboratories	12–30 m min^{-1}
Type C	Fume cupboards for special requirements involving toxic hazards	30–60 m min^{-1}
Type D	Movable fume cupboards and bench hoods	As for Type A, B, or C depending on requirements

*Reproduced by permission of British Standards Institution, 2 Park Street, London W1A 2BS, from whom copies can be obtained.

Fig. 1.7. A typical fume cupboard arrangement. The inlets which automatically come into operation to supply the additional air when the fans are switched on are above the cupboard. The right-hand cupboard houses a muffle furnace. All the services have the controls at the front and the two left centre compartments have a sliding division to provide a wide working area if required.

the points discussed above. Certain additional special features are shown, e.g. the left-hand cupboard houses a cooling bath, the centre two can be used as such, or the middle bar can be raised so that a large area is made available for work and there are two compartments at the right-hand side, one of which houses a furnace. Each group of cupboards is vented separately and the flow to each cupboard is adjusted by an internal baffle. The framework is beech and the walls asbestolite, both being painted with polyurethane paint. All the controls are at the front and are carefully colour-coded. The electrical sockets are at the rear of the cupboard and the switches at the front.

Fume cupboard ducts and fans

There are three types of extraction system in general use. The axial flow and propeller fans are capable of moving large volumes of air cheaply but it is necessary to have short ducting runs as they cannot work against any appreciable resistance. It is preferable for the motor to be mounted outside the gas stream but if this cannot be done then both the motor and propeller have to be protected against corrosion by painting or spraying with a plastic finish. A flameproof motor must be fitted if flammable vapours are likely to be present.

The most widely used fans are of the centrifugal type and are made in a variety of sizes. The motor is mounted outside the duct carrying the fumes and only the fan blades have to be protected against corrosion. For really

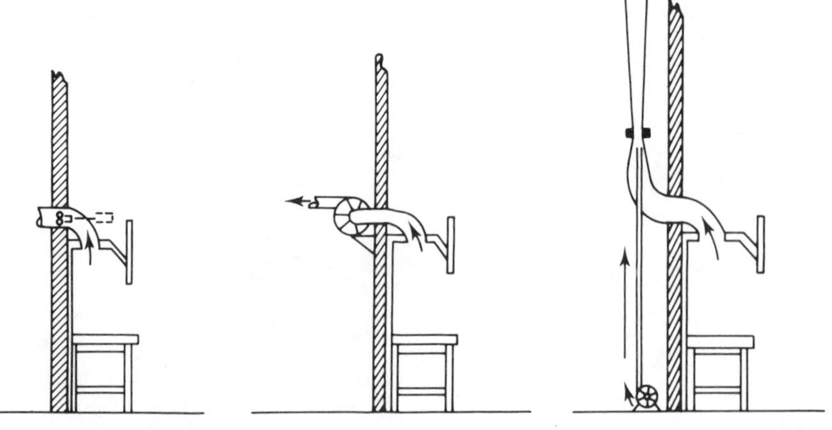

Fig. 1.8. Fume cupboard extractor systems: (a) axial flow and propeller fan; (b) centrifugal fan; (c) ejector fan. (Material from BS 3202, 1959 is reproduced by permission of the British Standards Institution, 2 Park Street, London W1A 2BS from whom complete copies can be obtained).

severe corrosion conditions the ejector type of fume extraction is used (Fig. 1.8). This works on the venturi principle, i.e. a jet of steam or compressed air creating a suction effect which draws gases from the fume cupboards. As the motor and fan or nozzle are outside the corrosive gas stream they can be made in mild steel. This type of evacuation system can be noisy unless steps are taken to soundproof the jet.

The ducting should be made as short and direct as possible and may be made from rigid PVC or polythene, galvanized mild steel with a suitable protective coating, asbestos cement coated with a protective material such as chlorinated rubber paint, or chemical stoneware. The duct exit must be positioned away from the ventilation intake to prevent the fumes being carried back into the building. A careful study should be made of the air movements around the building and the duct exit positioned so that it is clear of the air intake to the ventilation system, open windows and doors, and any other areas where people may be present. A high velocity nozzle can be fitted to the duct to give a long throw to the exit gases. A trap should

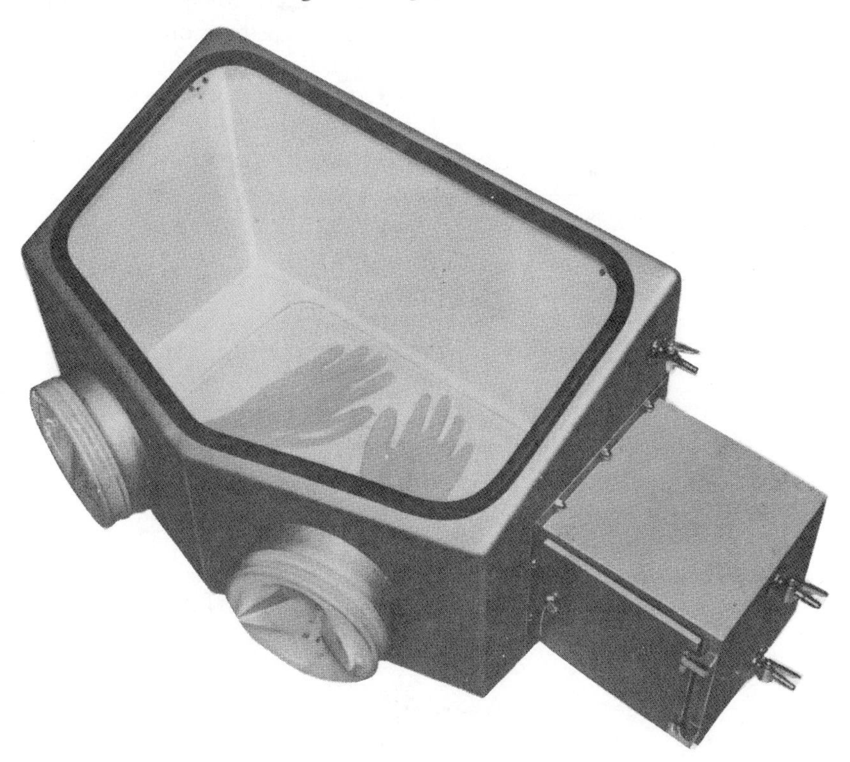

Fig. 1.9. Glove box showing inlet and outlet valves, airlock and handling arrangements. (Photograph courtesy of A. Gallenkamp & Co. Ltd).

be installed in the duct to collect any liquid which may drain from the system.

Fume Extraction Enclosures

In a previous section the desirability of installing a dedicated fume extraction system for a single experiment was mentioned. Such systems consist of

Fig. 1.10. A fire point equipped with fire alarm, fire extinguishers, fire blanket, water hose, and equipment for obtaining access to the roof space and drainage system.

small hoods which are placed over the exit from the experiment and they may or may not incorporate a gas scrubber to remove the objectionable constituents of the gas. Relatively high rates of extraction of the order of 50 m min^{-1} and draught-free conditions are desirable.

For work involving the handling of radioactive, bacteriological, or toxic materials it may be desirable to use a glove box (Fig. 1.9). These can be made in a wide range of materials such as mild or stainless steel, plastics, toughened glass, or fibre glass. Glass windows are inserted so that it is possible to see inside. The material should have a smooth finish to facilitate cleaning. The cabinet is equipped with (a) inlet and outlet valves so that the atmosphere inside can be exhausted and replaced, together in some cases with in-line filtration equipment; (b) an air-lock of reasonably large dimensions so that materials and apparatus can be introduced into the box without disturbing its atmosphere or releasing gases to the laboratory; (c) suitable connections for introducing services into the box; and (d) two or more glove ports with attached long-sleeved rubber gloves to permit working inside the box. An alternative system for passing material out of the box is to use a plastic bag. The bag is attached to a port and material from the box is transferred to the end of the bag where it is isolated by two adjacent rows of welding across the bag. The material is then removed by cutting the bag between the two rows of welding. Materials can be passed into the box by reversing the process. These boxes are particularly useful for handling small amounts of materials of high hazard rating as they can be confined to a small area of the laboratory. Needless to say, an explosion or fire in a glove box would have serious consequences and if working with flammable materials all sources of ignition must be removed, and preferably an inert atmosphere should be maintained.

Where work is carried out on a larger scale, specially designed cabinets may be used. Examples of these are walk-in fume cupboards in which the working area extends from the floor to the ceiling, radiation working areas equipped with remote handling equipment, and animal exposure cabinets. Hughes[3] has summarized progress made in the design and use of these units.

SAFETY EQUIPMENT

In a building housing a number of separate laboratories it is advisable to locate general safety equipment in an identical position in each laboratory so that members of staff will know where the equipment is, irrespective of the place where they are working. The most suitable place is in the entrance lobby or adjacent to the entrance door and the equipment should comprise: telephone with emergency numbers prominently displayed, fire-fighting equipment appropriate to the hazards in the laboratory, fire blanket, protective clothing, eyewash bottle, respirators and/or compressed breathing equipment, and antidote cabinet. In the case of work where the hazard rating is high, then the necessary safety equipment should be placed near the bench during the course of the work.

Fire protection is dealt with in detail in another chapter. Consideration must be given to it at the design stage, of course. It is cheaper and easier to install fire points at the initial construction stage than later. An illustration of one type of fire point is shown in Fig. 1.10.

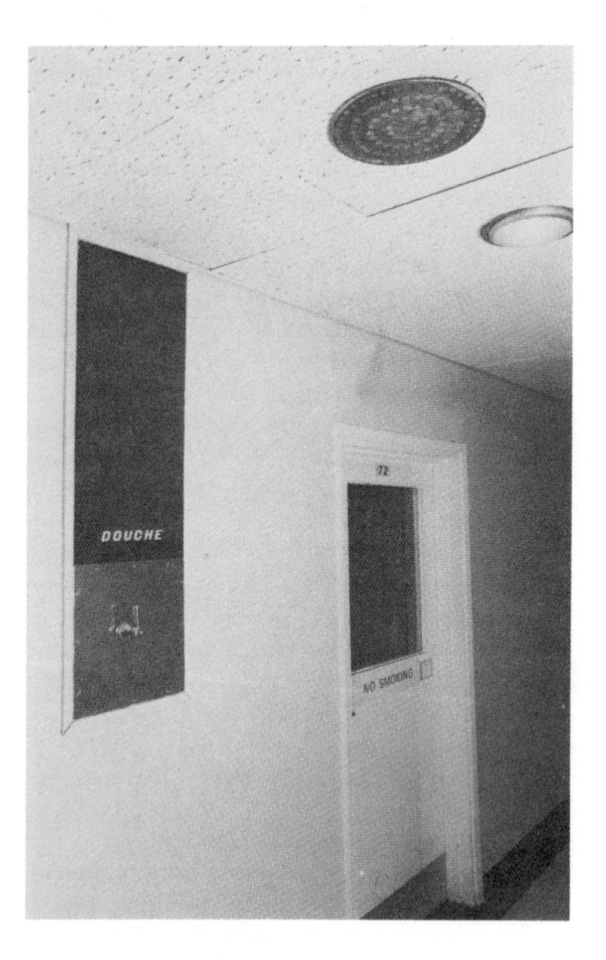

Fig. 1.11. An emergency drench shower situated over a drain in a corridor.

Safety Showers

These too are dealt with elsewhere. At the design stage decisions need to be made concerning the type to be used and locations. In laboratories where the quantity of corrosives handled is large a shower in each is justifiable. Elsewhere corridor installations may be adequate (Figure 1.11).

Eyewash Units

These should preferably be built into the laboratory furniture. A typical installation is shown in Fig. 1.12. A piece of rubber tubing attached to a cold water tap so that a stream of water can be directed to a particular point is also a simple but effective addition to safety arrangements. Should this be

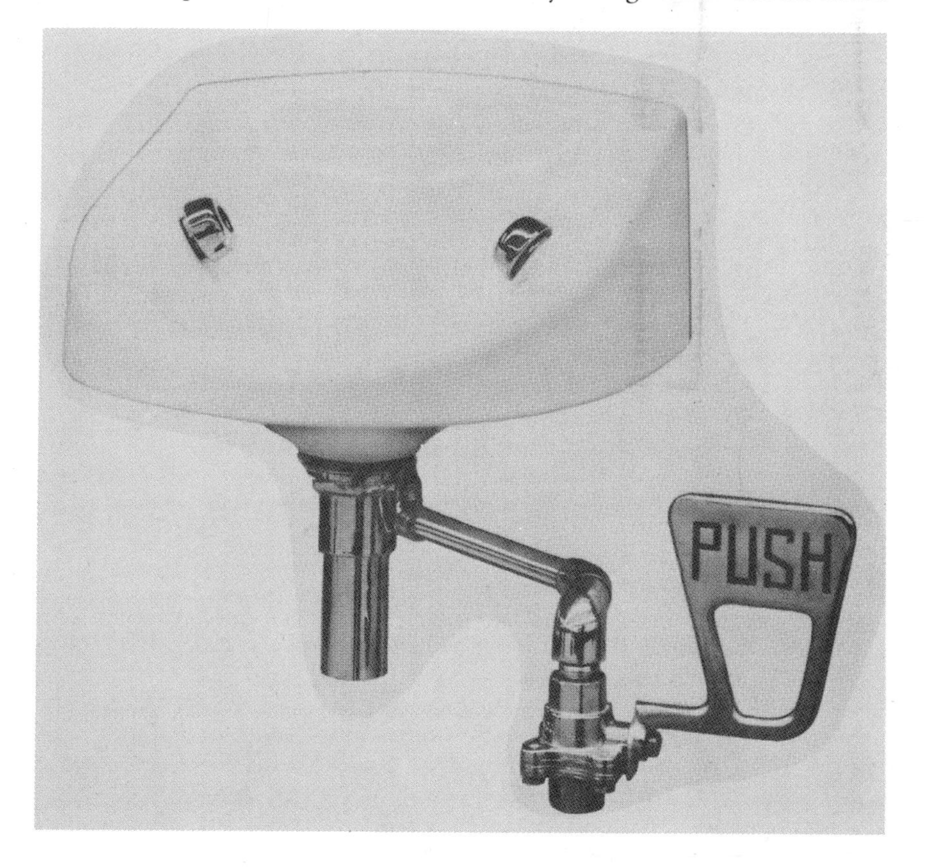

Fig. 1.12. A combined eyewash unit and drinking water supply. The water stream is aerated to provide a copious flow without undue pressure. (Photograph courtesy Walter Page (Safeways) Ltd).

used for eye irrigation the stream must be directed on to the nose and not directly into the eye.

REFERENCES

1. "Practical Laboratory Planning" (1973), Ferguson, W. R., Applied Science Publishers Ltd, London.
2. The Poisons Rules 1978 No. 1, Schedule 1, HMSO, London.
3. "A Literature Survey and Design Study of Fume Cupboards and Fume-dispersal Systems" (Nov. 1980), Hughes, D., Science Reviews Ltd, London. (Obtainable from 28 High Ash Drive, Leeds, UK, ISBN 0905927-50-8.)

FURTHER READING

Alderson, R. H. (1975). "Design of the Electron Microscope Laboratory", North-Holland Publishing Company, Amsterdam. American Elsevier Publishing Co. Inc., New York.

Boursnell, J. C. (1958). "Safety Techniques for Radioactive Tracers", Cambridge University Press.

Clark, R. P. (1980). The evaluation of open-fronted biological "safety" cabinets, *Laboratory Practice* **29,** No. 9, 926.

Cooke, J. (1979). Perchloric acid, the dangers of contamination in the laboratory, *Health & Safety at Work*, Dec. 1979, **54.**

Ellis, J. G., and Riches, N. J. (1978. "Safety and Laboratory Practice", The Macmillan Press Ltd, London.

Everett, K. (1966). "University of Leeds Safety Handbook", University of Leeds Safety Committee.

Ferguson, W. R. (1973). "Practical Laboratory Planning", Applied Science Publishers Ltd, London.

Grover, F. and Wallace, P. (1979). "Laboratory Organisation and Management", Butterworths, London.

"Guide for Safety in the Chemical Laboratory" (1972), 2nd edn. Manufacturing Chemists Association, Van Nostrand Reinhold Company, New York, Cincinnati, Toronto, London, Melbourne.

Guy, K. (1962). "Laboratory Organisation and Administration", Macmillan, London.

Hackett, W. J. and Robbins, G. P. (1979). "Safety Science for Technicians", Longman, London and New York.

Hooper, E. (1980). "The Safe Use of Electricity", The Royal Society for the Prevention of Accidents, Purley, Surrey.

Hughes, D. (1974). "The Design and Installation of Efficient Fume Cupboards", *Brit. J. of Radiology*, **47,** 888–892.

Hughes, D. (Nov. 1980). "A Literature Survey and Design Study of Fume Cupboards and Fume-Dispersal Systems", Science Reviews Ltd, London, ISBN 0905927-50-8. (Obtainable from 28 High Ash Drive, Leeds, UK.)

"Identification Colours for Pipes Conveying Fluids in Liquid or Gaseous Condition in Land Installations and on Board Ships" (1966). ISO recommendation R 508. International Organization for Standardization.

"Living with Radiation" (1973). National Radiological Protection Board, HMSO, London.

Mann, C. A. (1969). Safety in the chemical laboratory LVIII, science experiment safety in the elementary school, *J. Chem. Educ.* **46,** Part 5, A347–353.

Paget, G. E., ed. (1979). "Topics in Toxicology—Good Laboratory Practice", MTP Press Ltd, London.

Pieters, H. A. J. and Creyghton, J. W. (1975). "Safety in the Chemical Laboratory", 2nd edn., Butterworths, London.

"Radiological Protection in the Universities" (1966). The Association of Commonwealth Universities, 36 Gordon Square, London, WC1.

"Recommendations on Laboratory Furniture and Fittings", BS 3202 (1959), British Standards Institution, London.

"Safety Requirements for Fume Cupboards, Performance Testing, Recommendations on Installation and Use", BSI draft document 79/52625DC, British Standards Institution, London.

Scott, R. B. and Hazar, A. S. (1978). Liability in the academic chemistry laboratory, *J. Chem. Educ.* **55,** Part 4, A196–A198.

"Specification for Identification of Pipelines" (1975). BS 1710, British Standard Institution, London.

Stalzer, R. F., Martin, J. R. and Railing, W. E. (1967). Safety in the laboratory. *In* "Treatise on Analytical Chemistry", Part III, Vol. 1 (I. M. Kolthoff, P. J. Elving, and F. H. Stross eds), Interscience Publishers, New York, London, and Sydney.

"The Storage of Highly Flammable Liquids", *Chemical Safety/2* January 1977, Health and Safety Executive, HMSO, London.

Steere, N. V. (1971). "Handbook of Laboratory Safety", C.R.C. Press Inc., Boca Raton, Florida, USA.

Young, J. R. (1971). "Responsibility for a Safe High School Laboratory", *J. Chem. Educ.* **48,** A349–A356.

Young, J. R. (1970). Survey of safety in high school chemical laboratories of Illinois, *J. Chem. Educ.* **47,** Part 12, A829–838.

2

Organizing for Safety

When Shackerly Marmion postulated that familiarity begat boldness[1] he could well have had laboratory workers in mind. Scientists become preoccupied with their work and overlook obvious hazards. Equipment is used without mishap for long periods, and normal sensible precautions are neglected. The diversity and complexity of equipment and apparatus may be so great that dangers are forgotten. These things are understandable, even though they cannot be condoned. But it is not only these quirks of human nature which have to be considered and controlled.

A fact of life in the latter half of the twentieth century is that people's habits and outlook on work have changed. Unacceptable though it may be in many ways it is irreversible. This does not mean that we should throw up our hands in despair and accept standards in laboratories which are manifestations of this change. It means that we should acknowledge personality changes as a fact of life and do all that lies in our power to counteract them. To ignore them in the hope that they will go away or that there will be an automatic return to the habits of a few decades ago, is a negation of safety.

These changes reveal themselves at the lowest levels—technical assistants, laboratory assistants, people taken from process work in factories and trained in simple analytical techniques, and so on. Like the creeping branches of a malignant growth, bad laboratory habits spread upwards to the higher echelons unless checked at the source. Even some graduate chemists are guilty of unsafe practices. Many are unaware that there is another and better way of working, and cannot understand what they consider to be unnecessary fuss. Educational establishments have a part to play here. Unless in schools and colleges there is rigid insistence on good laboratory practice, the observance of high standards of personal discipline and that imposed by others in authority, the habits thus developed will continue to be practised in working life. This is not a blanket condemnation of school laboratories. Teachers and lecturers have the same problems as employers in trying to impose reasonable but essential standards. In their cases, those on the receiving end are more malleable and receptive (even if in a changing world not always so); they have not acquired bad habits needing change. Their role is therefore the more important.

PERSONAL CLEANLINESS

Standards of personal hygiene need to be even higher in laboratories than most other work situations. Dermatitis is a generic word applied to any inflammation of the skin. Primary irritation dermatitis can be produced by mechanical agents such as friction, and from chemical agents like irritant gases and vapours, acids, and alkalis. If exposed parts of the skin are thoroughly washed as soon as possible after contact occurs then the possibility of skin trouble is substantially reduced. When meal or coffee breaks are taken it should be accepted practice to wash hands first, irrespective of whether there has been chemical contact. Work in high temperatures is common. Showers should be used when such work ceases or at the end of the work period. Particular attention should be given to underarm and crutch areas. Dirty outer clothing and footwear should be changed and cleaned as soon as possible after contamination occurs. Besides aesthetic, health, and safety reasons for high standards of personal hygiene, almost always there are also practical considerations related to the work in hand. Soap and hot water are the best cleansing agents. Acetone, benzene, and other organic solvents will remove certain stains, but also promote dryness of the skin. They are not to be used.

FOOD AND DRINK

This is a contentious matter. Tea and coffee are sometimes made in laboratories amid toxic and corrosive chemicals and actually drunk from beakers. Ideally no food or drink should be allowed in laboratories or chemical stores. It is difficult to secure compliance with this rule if no proper facilities are conveniently located nearby. In such cases first try to rearrange the area so that facilities can be provided. Only as a last resort should a small area within a laboratory be set aside for food and drink. It must be isolated from the work areas by fixed screens and strict control exercised to make sure that no harmful materials encroach on the area. While not wholly desirable, this arrangement is better than the indiscriminate consumption which would otherwise occur. Where communal eating facilities are used protective garments should be removed before canteens are entered. Apart from the personal risk from contaminated clothing and footwear, soils can be passed on to seats and floors and so be transmitted to others.

HOUSEKEEPING

Failure to clean up is a notable manifestation of declining standards. Every laboratory worker should automatically clean up at each stage of work and, finally, at the end of a work period. There is now a tendency to expect others to do this. Cleaners do not know enough about the nature of spillages to deal with them safely and should not be expected to do so. Sometimes no cleaning at all is done apart from an occasional brush over the floor. Laboratory equipment, bottles of chemicals and reagents, often covered with dust, clutter up benches so that the next user has to push the accumulation into an even more untidy heap to create a little space. An essential element of induction training should be instruction in the cleaning procedures expected. Thereafter it is a question of supervisors and managers insisting on the maintenance of standards. A training programme should embrace the following housekeeping elements:

1. Immediate mopping up of any spillages on bench or floor, taking into account the chemical spilled, and the safe disposal of cleaning materials.
2. Keeping the bench clean from chemicals and apparatus other than those in immediate use.
3. Cleaning up after each stage of an experiment.
4. Keeping floors unobstructed, dry, and free from slippery materials.
5. Returning apparatus to its proper storage space in a clean condition ready for the next user. Arranging repair needed before return.
6. Cleaning ring stands and tubing before being left. Residues may cause irritation or burns to the hands of the next user.
7. When apparatus containing corrosives cannot be washed immediately, rinsing out before leaving in sinks for final cleaning.
8. Keeping floors clear of reagent bottles and apparatus.
9. Returning reagents and chemicals to their proper shelves after use.
10. Keeping labels to the front.
11. Cleaning off spillages from the side of bottles.

SMOKING

Rules on smoking may vary according to the kind of work. It is undesirable in any circumstances and in laboratories is accompanied by special risks to people, apparatus, and even buildings. A complete ban on smoking should be the goal. If safe areas for smoking are provided away from working laboratories but close enough to be convenient, no smoking rules are more likely to be obeyed. Good example and meaningful disciplinary action where breaches occur also help. It is unwise to forbid smoking if there is a doubt as to whether the ban can be enforced. Whatever the situation smoking cannot be allowed in some circumstances, such as where there are chlorinated hydrocarbons, fluoropolymers, other flammable liquids, laser areas, and around unsealed sources.

WORKING ALONE

Sometimes laboratory people may have to work alone. In small organizations there may be only one chemist or assistant. In such cases there must be facilities for raising the alarm if help is needed. Sophisticated equipment can be used. One example is a pocket device which sends an alarm signal to a control point if the wearer falls. There is a time lag between a fall and the despatch of the signal to take account of accidental falls. If the wearer regains his feet within 10–15 seconds the alarm does not operate. The other extreme is a simple wire or cord stretching perpendicularly above benches. This can be seized if dizziness, faintness or possibly intruder attack occurs. An alarm system is activated, varying between a simple hand bell or electric bell(s). The fact and circumstances of people working alone must be known. Chemists and others may do so as a matter of routine, in which case failure to be seen or make contact from time to time will be noticed. Those who occasionally work alone should make others present in the building aware. Some may come in outside normal hours or stay back after others have finished. Checks can be organized when this happens. Janitors, cleaners, watchmen, and so on can make occasional visits. A system can be organized for periodic telephone calls to manned locations, possibly a gatehouse or even the home of a duty manager or similar person. An every half-hour outward call alternating with a half-hour inward call is reasonable. This system is effective when a building is otherwise unoccupied.

Reasonableness enters into any of these arrangements. Where work is in a completely safe area where innocuous materials are handled contacts are less important than when the reverse is the case. In an ideal world, working alone would be avoided. The practicalities of life indicate that it is sometimes necessary.

RUBBISH DISPOSAL

A method for disposal of waste other than chemicals in a large laboratory block was for cleaners to push wheeled metal containers to the door of each laboratory. Rubbish and harmless waste was separated from broken glassware in different bins, each of which was tipped in turn into the appropriate wheeled containers, which when full were taken to a safe compound where the contents were collected the next morning by the local authority refuse collection service. The collectors were aware of the nature of the waste and suitable handling and eventual disposal methods were devised and operated. However, on one occasion a cleaner was tipping the contents of a broken glassware bin when liquor which had been left in part of a broken bottle ran down her arm. It was concentrated sulphuric acid, she suffered severe burns to one hand and forearm, was unable to work for many months and eventually secured a substantial compensation payment. An unnecessary and serious accident resulted from thoughtless disposal of broken glassware. Residual matter should be safely disposed of and rubbish bin covers kept closed. Overflowing bins are a source of possible contamination, and fire hazard, and a sign of poor attitudes to housekeeping. Regular cleaning of interiors should be a matter of routine.

POISONS

Occupational poisoning is fortunately rare, but poisoning as a cause of death is not. In any one year, in the United Kingdom about the same number of people will die as a result of poisoning as from road accidents. In the United States almost 75 000 poisoning cases will occur, almost half of which will be children. Worldwide the majority of poison deaths are classified as suicide, and two-thirds of the deaths from this cause in the UK are due to self-administered toxic substances.[2] Laboratory workers, surrounded as they are by poisons, are often less careful than they should be in handling and storing them. Although deaths from industrial poisoning are fairly infrequent the high incidence of poisoning generally indicates that special care is needed to ensure that toxic substances do not get into the wrong hands.

Any substance which carries the international poison symbol should be treated as hazardous. This seemingly obvious statement is made because careless handling of poisons is often excused by the comment "although it is marked POISON it's harmless". Arbitrary decisions of this kind cannot be left to individuals. There should be a secure, lockable cupboard for poison storage with prearranged key control. The amount of poisons out on benches and shelves should be the minimum compatible with efficient operation. Containers should be placed so that they cannot be confused with other substances. For instance, bottles containing poisons have been observed adjacent to eyewash bottles. The fact that eyewash should be in proper, easily accessible cabinets is irrelevant; the juxtaposition is wrong anyway.

LABELLING

Chemicals should be clearly labelled and carry the appropriate symbol. Altering or tampering with labels should not be allowed. There is a simple procedure which must be strictly adhered to when containers are to be used for substances other than the original contents. First the container should be correctly cleaned and dried, the original label removed by soaking, then a clean label fixed showing the new contents. Crossing out labels and writing on them is inadequate and hazardous. Some countries have statutory labelling requirements for named chemicals which must be followed where applicable. If there is any doubt as to the contents of containers which cannot be positively identified because of missing or illegible labels, they should be carefully and separately disposed of in accordance with rules set down elsewhere. Liquids poured in a direction away from the label will avoid damaging it. Standard solutions should carry the date of preparation and initials of the person making them up. In most countries labels complying with local legislative requirements can be purchased from an approved source.

In summary, labels should bear the following:

1. Name of product.
2. Any mandatory symbol, pictogram or other matter.
3. If not included in 2, the degree of hazard and associated information, e.g. danger, warning, caution.
4. Precautionary measures criteria, properties, and use.

Labelling can be an important means of training. Users cannot avoid a hazard unless they know about it. Informative labels, used as a discussion topic in training sessions, help to raise the general level of safety.

SECURITY AND SAFETY CHECKS

Safeguarding human life, buildings, and property are prime functions of security. Where laboratories are concerned there are additional special considerations. Individuals or groups of people may be antagonistic towards the kind of work carried on. Some substances used or stored are sought for their narcotic or stimulant effect; platinum crucibles and mercury are valuable. The consequences of fires deliberately started by intruders—a common facet of modern life—extend way beyond the intrinsic value of buildings and equipment, since records and reports concerning development work over a long period may be destroyed. The need for comprehensive security arrangements is greater in such establishments than in most others.

Laboratories contained within the curtilage of a factory or other enterprise can be included in the general security arrangements for the site as a whole. Isolated laboratory blocks are a special case. Whatever the circumstances a security plan tailored to the particular needs of the location must be devised. Sizeable installations will justify a 24-hour watchman system. Others, because they are part of a complex, can be the subject of periodic checks from persons working elsewhere on the site. A third possibility is the use of guards from a security organization, on a spot check or full or partial cover basis.

The 24-hour watchman cover must be sensibly used to give proper checks. Reliance on conscientious surveys at a reasonable frequency unsupported by mechanical back-up is unsatisfactory. Human failings account for a high percentage of accidents, and so too do they account for breakdowns in security. In isolation, during the hours of darkness in particular, even the most alert guards can fall asleep or forget. Back-up systems exist which take account of these fallibilities. One example is an installation linked to the nearest police headquarters. At random times a bell, loud enough to wake the acute somnambulist, rings to warn that a security check/tour must begin. The watchman tours all parts of the laboratories, inserting a key and turning it at preselected checkpoints. Should he fail to return to his starting-point, having operated all the check mechanisms, within a predetermined time an alarm signal is automatically transmitted to the police.

The telephone can be used in place of, or to supplement, such a system. Telephone contact, also at prearranged times, is made with a continuously manned point, possibly the office of a security organization, another site where a watchman is on duty, or another location of the same firm where shift work is carried on. Enterprising companies can form consortiums where this is geographically feasible. Whatever the arrangement, when a call is not received at an agreed time investigations can begin.

Security checks should not only take account of possible intruders. Sometimes it is necessary to leave apparatus running outside normal hours. Guards should be informed in writing when this happens. Detail should include location of apparatus, frequency of checks, abnormalities to look for, action to be taken if things go wrong, and the telephone number of the person to contact in emergency. A check list prepared in advance will help to ensure that simple but important aspects of security are not overlooked. Here is an example:

1. *Guards or watchmen*
   a) Is there a surveillance system covering a 24-hour cycle involving
      i. full-time guards, or watchmen,
      ii. occasional site visits through a security organization,

iii. occasional checks outside normal working hours through employees?

b) Are guards properly trained and is there a checking system to ensure that security duties are properly carried out?

c) Are there electronic or other back-up facilities for guards to secure assistance when needed?

d) Is communications equipment in a secure area with adequate protection against sabotage for wires and cables?

2. *Protection of documentation etc.*

a) Is there duplication of computer discs, data files, etc. with copies in a secure store away from the laboratories?

b) Is the computer facility in a secure location, with automatic protection approved by the manufacturer and regularly serviced and tested?

c) Is other important documentation safely stored in fireproof safes or cabinets? Is it possible for unauthorized photocopies to be made?

d) Are confidential files etc. safely locked away as necessary?

e) Is the system for disposal of confidential documents foolproof? If carbon ribbons are used on typewriters, are they of the self-effacing type, or are they destroyed immediately after use?

3. *Fire protection*

a) Are guards familiar with extinguishers and their location, fixed hose reels, etc.?

b) Is the emergency procedure known and practised?

c) Have all reasonable steps been taken, having regard to the extent of the risk, in relation to compartmentation, fire and smoke stop doors, roofs and other areas?

d) Are flammable liquids safely put away in flameproof cabinets, or in approved stores?

e) Has the advice on fire prevention and protection such as outlined in Chapter 9 of this book been studied and applied where possible?

4. *Perimeter security*

a) Is there a fence protecting the site which is

i. high enough, protected at the top to make egress more difficult, so designed that it is not possible to crawl under, in good repair,

ii. are permanent structures such as small storage buildings, incinerators, etc. far enough away so they cannot be used in scaling the fence,

iii. are moveable objects like ladders, rubbish bins, etc. lying around and accessible for use in climbing the fence?

b) Is there reasonable, not repressive, gate security? Are the gates in good repair, and of adequate strength?

c) Is artificial lighting adequate to reveal intruders?

d) Are rubbish compounds secure, are rubbish or other flammables easily accessible for incendiary purposes, is the clearing and collection of rubbish adequate?

5. *Strong-rooms, safes, etc.*

a) If a strong-room or vault is justified, so too must be automatic protection systems operating when any unauthorized opening procedure is attempted.

b) Are safes fire resistant, securely anchored to wall or floor, sited so that surveillance from outside is possible, and not fixed to a wall having easy and unobserved access on other side?

6. *Doors and other openings*

a) Are doors and frames constructed to a degree of strength and security related to the potential risk. Are external doors secured by deadlocks, or bolts integrated into doors and drawn only by a key? If there are glass panels are they protected where necessary by bars or expanded metal and, if broken, can the door lock be reached?

b) Are these design features present:

i. lock cannot be prised open,

ii. hinges so constructed and sited that they cannot be broken,

iii. bolts located so they cannot be cut, locks mounted so that they cannot be prised off and will remain secure if the frame is split?

c) Is there a check to see all doors and windows are properly secured outside working hours, and is key security adequate?

d) Are gratings and bars placed over easily accessible windows? Is the number of openable windows at the minimum compatible with environmental needs? Has a check been made to see if it is possible to replace some windows on lower floors by glass blocks, or reglaze them with high impact resistant plastic? Are windows adjacent to external fire escapes protected? Are window locks placed so that it is difficult to reach them if the glass is broken?

e) Where there are skylights, can they be damaged by thrown objects, are they protected with bars if vulnerable, are roof hatches secure? Has the need for any, or all, roof openings ever been questioned?

These suggested components of a check list can be used to build up a list made up of such of the items as are applicable to each location.

PERSUADING PEOPLE TO WORK SAFELY

"Unsafe acts of persons . . . are the direct and proximate causes of the majority of accidents" wrote Heinrich,[3] the high priest of accident prevention. Internal studies carried out in the United Kingdom by British Gas Corporation and others have identified around 85 per cent of all industrial accidents as having their origin in unsafe acts. These unsafe acts need not necessarily be on the part of an injured person or accident victim. Somewhere in the chain of events ending in an accident someone has committed an unsafe act or omission: an instruction has been poorly expressed or misunderstood; training has been unsatisfactory; the wrong tools or equipment have been chosen; people have been assigned tasks beyond their capability; someone has decided to carry out work in a way other than that laid down, and so on. Therefore if accident prevention activities are confined to the hardware of laboratory operations—the buildings, apparatus, equipment, substances, etc.—then the attack is being directed to areas where the effect is likely to be less rewarding than concentration on improving human behaviour. Just as it would be wrong to ignore the physical aspects of prevention so too is it even more undesirable to ignore the "people" aspect. Yet concentration is almost always on the former.

If it were possible to achieve a state of affairs in which everyone from the director or senior manager down to the laboratory janitor worked together at maximum levels of efficiency and safety (the terms are really synonymous) then the results would be staggering in terms of output standards and freedom from accidents. But it is not possible to achieve this in totality. Every working day a number of people will assemble in laboratories and remain in more or less close proximity for the whole day. Each is an individual with characteristics, capabilities, and views on life which will differ to a greater or lesser degree, leading to compatibility or incompatibility with their peers. It is too much to hope that a completely harmonious togetherness will result. Still, there are factors common to all on which it is possible to work to direct them to collective goals. Five years work on behavioural problems in the Hawthorne Plant of the Western Electric Company[4] carried out by Elton Mayo showed that the use of motivational techniques commonly regarded as successful were not necessarily so; the introduction of advantageous payment systems, rest periods, free meals, and similar improvements in working conditions resulted in the expected improvements in efficiency. When these benefits were taken away efficiency rocketed to its highest level ever, which nonplussed Mayo and his colleagues. The eventual conclusions were that participation in the experiments had created a unification of purpose within the group involved. Hence Mayo propounds that morale is more important than the physical conditions of work.

Frederick Herzberg set down ideas in 1968 which still hold good.[5] His activators he identifies as achievement, recognition, work itself, responsibility, and professional growth. Some kind of external pressures must be applied if people are to do their jobs well and Herzberg believes that without them morale is poor. More closely related to the laboratory worker is the six-year study of M. Scott Myers.[6] He studied the job motivators and dissatisfiers of scientists, among others. Achievement was the factor identified most often, and more than twice as many interviewees considered its acknowledgement as a favourable motivator than otherwise. Recognition as an individual ranked next in importance.

There are almost as many views on factors which influence the way in which people work as there are writers on the subject. Nevertheless there are certain combinations common to most. Included in these are high morale, and recognition of achievement, and the individual. Personal experience has shown these to be powerful motivators towards high levels of safe working. The key word is involvement. Every channel of involvement should be exploited. Beginning with the initial interview of prospective employees it should be made clear that laboratory safety is regarded as having prime importance. It is an element taken into account in individual assessments and advancements. Achievement is measured in terms of quality, output, and safety. Recognition is given not to laboratory workers who attain in simple terms high levels of unit output where this is measurable, but to those who maintain acceptable levels of standards, performance, and safety. An induction programme should contain a substantial element of health and safety training. The employer's goals in these areas should be clearly spelled out and acknowledged.

Thereafter there should be no barriers between individuals and the attainment of these goals. Poor example on the part of senior managers, the toleration of faulty or unsafe equipment, the acceptance of low standards of personal and collective hygiene are examples of barriers which will induce frustration and lower morale. The converse is the building of bridges between managers, supervisors, and others. Traditional bridge-building methods are general and specific: safety committees, calling together at regular intervals all those in a laboratory who are free at that moment for discussion of a safety topic preferably related to work in progress, face-to-face discussions on health and safety aspects of a current task and dissemination of information about incidents within or outside the place of work. Managers sometimes say they do not have time to carry out any or all of these activities. Investigation usually shows that it is the will, not the time, which is missing.

Everyone in a laboratory should be aware of safety targets set both for their own location and the site or organization as a whole in large or multi-unit operations. The achievement or otherwise of these targets should be made quite clear. A sure test of the extent of employee involvement is to ask for information regarding the safety performance in a laboratory. If the reply is vague or uncertain then there is a shortfall in involvement. Targets should be realistic and not, for example, limited to the attainment of a given number of accident-free hours. For instance, rules regarding the wearing of eye protection may not be adequately accepted in one area. A target could be the raising of the standard to the required level within a specified time. Whatever the aim, discussion followed by agreement as to its desirability is an important prerequisite to its final establishment. When success comes there must be suitable acknowledgement. Individual actions revealing a special involvement should receive recognition. Uplift in morale is an inseparable accompaniment of involvement and its maintenance is a continuing process. When the involvement machinery starts rolling constant refuelling will ensure that it does not stop.

REFERENCES

1. "The Antiquary", Act 1, Sir Walter Scott.

2. Encyclopaedia of Occupational Health and Safety, Vol. II (1972). International Labour Organisation.
3. Heinrich, H. W. (1959). "Industrial Accident Prevention", McGraw-Hill, New York.
4. Mayo, E. "The Social Problems of an Industrial Civilization", Harvard University Press, Cambridge, Mass.
5. "How do you activate employees?" *Harvard Business Review*, Jan./Feb. 1968.
6. "Who are your satisfied workers?" *Harvard Business Review*.

FURTHER READING

Stevenson, A. "A Guide to Action—Planned Safety Management", Alan Osborne and Associates (Books) Ltd. Unit 5, Seager Buildings, Brookmill Road, London.

Hughes, E. W. "Human Relations in Management", Pergamon Press, Oxford.

3

Hazards of Laboratory Equipment

The modern laboratory with its range of sophisticated electronic equipment has changed radically in the last few decades from just being a place in which most of the work is carried out in glass equipment using gas-fired burners for heating purposes. Nevertheless these basic techniques and apparatus are still used in laboratories, and accidents involving them continue to occur because of lack of knowledge of the necessary safety precautions or because they are wilfully ignored. An examination of any accident treatment book in a general chemical analytical laboratory will almost certainly show that many accidents are caused by cuts or burns. It is important that any worker in a laboratory has a good knowledge of the precautions necessary when handling basic equipment and any training course in laboratory techniques should start at this stage.

These basic techniques are dealt with in this chapter. Specialized laboratory instrumentation such as X-ray spectrometers or lasers, and equipment such as roll mills and grinding equipment are dealt with in a later chapter.

GLASSWARE

Glass is a most useful material to the scientist. It can be worked into the most complicated shapes, it is resistant to attacks by almost all chemicals (the exception being fluorides and hydrofluoric acid), it is transparent and unaffected by heat up to fairly high temperatures, the limiting temperature being its melting point. Unfortunately, it suffers three major disadvantages: it is fragile, broken pieces are sharp and cause serious wounds, and it is susceptible to thermal shock. The latter, however, can be overcome by using pure silica which can be obtained in transparent form and which has a negligible coefficient of thermal expansion, but it is also much more expensive than glass. To reduce the effect of extreme temperature, glassware is made as thin as possible which increases its fragility. Because of this thickness, apparatus needs to be carefully supported. Large flasks and those containing heavy loads should be lifted with two hands, one of which is under the base of the flask. The flasks must not be supported by a retort clamp around the neck but should be placed on cork rings, isomantles, or suitably protected tripod stands. Flasks containing material should not be lifted by the sidearm. Beaker tongs are useful for holding beakers but they should not be used for large-sized beakers, i.e. greater than about 1 litre in size.

Graduated cylinders can tip over and break very easily. This can be prevented by placing a thick rubber disc round the top of the cylinder so that this makes contact with the bench before the cylinder if it is knocked over. Volumetric flasks are also fragile because of their long necks and these can also be protected using a similar type of rubber disc.

Many accidents are caused by the incorrect use of pipettes. It is good practice to always use a suction bulb irrespective of the type of liquid being pipetted. The modern devices are extremely efficient and give a fine degree of control so that precise measurement is possible. In no case must corrosive or poisonous liquids be pipetted by mouth. The filling of a burette is another operation which can result in injury; this is easily avoided by using a small funnel to facilitate pouring the liquid into the fairly narrow aperture of a burette. Self-filling burettes, in which the reagent bottles are housed at ground level, also reduce the possibility of an injury.

Pressure build-up can occur quite easily in separating funnels, particularly if a volatile liquid is being used. Pressure must be released frequently during the extraction process. Glassware which is used at low pressure must be protected to avoid injury if explosion occurs. Special wire cages are available for standard shape items such as desiccators, winchesters etc., and tough plastic film is obtainable to protect flasks and special apparatus.

When heating liquids in test-tubes, superheating can easily occur resulting in the violent ejection of hot liquid. The test-tube should be pointed away from the body or other persons nearby, and the liquid should be agitated gently by shaking the test-tube. Heat resisting tubes should be used for heating liquids.

Support of glassware in large assemblies must be carried out carefully, and due allowance must be made for the expansion of glass if the apparatus is to be heated. In the case of flasks containing large quantities of liquid it is good practice to place a tray underneath which is more than capable of holding the entire contents. This is important even in the case of innocuous liquids as the hot liquid can cause severe burns.

The relatively simple operation of inserting glass tubing into rubber bungs is responsible for an inordinately large number of accidents. The mistake is to bore too small a hole in the bung as a result of using a borer of the same size as the tubing and then to attempt to force the tubing into it. Breakage occurs very readily with resultant cuts of the hand if no protection has been used. A borer slightly larger than the tube or rod should be used, lubricated with soap or grease to facilitate cutting of the hole. The tube should be gently eased into the hole using a cloth or gloves to protect the hands. Alternatively the tube can be placed inside the borer while it is still in the bung and then the borer can be removed, leaving the tube in the bung. Corks should be rolled before they are bored and a borer similar in size to the tube should be used. The ends of tubing or rod should be fire-smoothed to avoid sharp edges. When removal of glass tubing from bungs or corks is difficult, insertion of a cork borer sometimes may be helpful. As a last resort the bung should be cut and split away from the tubing.

Before use, any glass apparatus which has been excessively heated or modified must be examined for signs of strain and if necessary reannealed. All glass apparatus before use should be examined for defect and if badly chipped or cracked must not be used. Stopcocks that have stuck in their barrels may be removed by the application of gentle heat or by lightly tapping the handle with the wooden end of a spatula with the thumb placed at the other side. Care should be exercised and gloves worn.

Finally it should be questioned whether glass equipment is necessary. The widespread availability of plastics and metal laboratory equipment now means that in many cases glass need not be used.

ELECTRICAL EQUIPMENT

The use of electrical equipment may range from simple apparatus such as hot-plates and ovens to complex instrumentation. In addition to the electrical hazards, such equipment may cause fires and explosions. The dangers of electricity are not necessarily associated only with power supplies; there have been many cases where serious damage has resulted from the effects of static electricity. Quite small voltages and currents have been known to cause fatalities and the quantitative effects on human beings are given in Table 2.1. Injury can also be caused by a small shock if the person receiving it is on a ladder or in a similar exposed position. The shock itself may not be harmful but it may induce muscle contraction which will cause the victim to release his hold. The amount of current passing through the human body depends on its resistance which, as is well known, is much reduced under wet conditions. The effect of the current depends on the path the current takes through the body. At normal mains voltage of 240 V 25–30 mA can cause fatalities and voltages as low as 70 V have been known to cause death. Electric shock victims require immediate treatment. If they are unable to let go because of muscular paralysis, the electricity supply must be switched off before they are touched otherwise the rescuer may also receive a shock. An approved method of artificial respiration must be

Table 3.1. Quantitative effects of electric current on man
*(energy in watt seconds).[1]

Effect	*Milliamperes*					
	Direct Current		Alternating Current 60 Hz		10 000 Hz	
	Men	*Women*	*Men*	*Women*	*Men*	*Women*
Slight sensation on hand	1	0.6	0.4	0.3	7	5
Perception threshold, median	5.2	3.5	1.1	0.7	12	8
Shock—not painful and muscular control not lost	9	6	1.8	1.2	17	11
Painful shock—muscular control lost by 0.5%	62	41	9	6	55	37
Painful shock—let-go threshold median	76	51	16	10.5	75	50
Painful and severe shock—breathing difficult, muscular control lost by 99.5%	90	60	23	15	94	63
Possible ventricular fibrillation:						
3-second shocks	500	500	675	675		
Short shocks (T in sec.)			$116\sqrt{T}$	$116\sqrt{T}$		
Capacitor discharges	50*	50*				

applied immediately and medical aid summoned. Resuscitation must be continued until the victim is revived or pronounced dead by a doctor. The victim usually receives burns which may be severe and these must be treated by qualified medical personnel. If the equipment in use in a laboratory is such that there is a greater possibility of accidental electrical shock then personnel should be trained in rescue procedure and, as they will be fully occupied in the event of an emergency, an automatic system should be installed for summoning outside help.

The guidelines given here are intended for the general laboratory in which no specialized work involving electricity *per se* is carried out. A qualified electrician must be insisted on when connecting equipment to the mains supply. The equipment should be carefully checked before current is switched on; this is particularly important in the case of home-built apparatus. Checks at frequent intervals—say every three months—should be made by a qualified electrician. Work on electrical equipment should be done only when the power supply has been isolated and precautions taken to ensure that it cannot accidentally be switched on again by withdrawing fuses or using lock-off switches. It is advisable to draw up a Code of Practice regarding electrical apparatus in which the actions and responsibilities of the electrician and the user of the equipment are carefully defined.

Code of Practice

The person in charge of the laboratory should:

1. know where the main isolation switch is located;
2. organize a check rota for the equipment;
3. ensure that cable lengths are reduced to the minimum to avoid hazardous trailing lengths;
4. ensure that all electrical contacts and conductors carrying potentially hazardous currents are enclosed;
5. keep highly flammable liquids away from electrical contacts;
6. ensure that the correct safety equipment is readily available and that staff in the laboratory know how to use it;
7. ensure that when two or more personnel are working on the same equipment, one person is delegated to be in charge and a system of operation is devised to prevent the equipment being switched on by one without the other's knowledge;
8. in cases where a flammable atmosphere may be present, ensure that electrical equipment is of the flameproof type.

These precautions are additional to those falling within the responsibility of the qualified electrician who installed or maintains the equipment.

Refrigerators and Ovens

Volatile flammable liquids can produce explosive mixtures with air at very low temperatures. When such liquids are stored in a refrigerator, to ensure complete safety, all electrical contacts, lights and thermostat switches should be absent from the compartment, which preferably should be vapour-tight. The liquid should also be kept in a closed container if possible. Any refrigerator not so treated should be clearly marked that it is unsuitable for the storage of flammable liquids.

Similar precautions should be taken with ovens which are used for drying materials containing flammable liquids. In the case of chemicals such as carbon disulphide which has an auto-ignition temperature of 100°C, purging with nitrogen reduces the risk of fire and an enclosed system is preferable, condensing and recovering the volatile liquid.

Electric Power Tools

Portable electric tools must have sound cables and plugs and be correctly earthed, unless they are of the double insulated type. Particular attention must be paid to the earth if the tools have to be used in damp or moist atmospheres. They should be checked periodically by a qualified electrician and a log kept of the service they have received.

Capacitors

Capacitors may hold their charge for a considerable time and they should always be shorted before being handled. It is advisable to label capacitors stating that they should be kept shorted when not in use. Where possible they should be coupled to discharge resistors to ensure that the circuit becomes completely dead shortly after the supply is removed.

Microwave and Radio Frequency Equipment

See section on pages 119–121.

Electrophoresis Equipment

Particular care should be taken when carrying out electrophoresis experiments as the use of aqueous buffer solutions and high voltages provide circumstances in which the risk of electric shock is high. The precautions necessary in this type of work have been reviewed by Spencer, Ingram, and Levinthal[1] who recommended complete enclosure of the equipment, the enclosures to be earthed and fitted with a microswitch so that the power is automatically isolated when it is opened. They also recommended that the equipment should be installed by a qualified electrician, safety devices should be duplicated or backed up by secondary devices to make it failsafe, and it should only be used by experienced personnel. Relatively low voltage units, i.e. those operating between 100 and 500 V, are commonly used on the laboratory bench and they should be clearly labelled with a sign to indicate that the power supply is switched on.

Electronic Equipment

Electronic equipment poses the hazards associated with all electrically powered apparatus, and it should be emphasized that only authorized persons carry out any work on it apart from normal operation. "Authorized persons" embraces only electronic engineers and qualified electricians. Even tasks such as the wiring of mains plugs and connections to the mains supply must be carried out by them and no one else. Twenty-two laboratory workers on a training course were handed mains plugs and a short length of three-core cable, then asked to wire them up: three connected wires wrongly. A salutary reminder was provided that such tasks are the province of the electrician. Wherever it is possible to do so low voltage equipment should be used in all electrical applications. Here are guidelines on precautions to be taken by those working with electronic equipment; they are not exclusive.

1. Equipment, whether or not of commercial manufacture, should be furnished with a circuit diagram, operating instructions, and an explanation of the associated hazards and

the operation of safety devices. Door interlocks should be installed (preferably in sight) to de-energize high-voltage circuits when doors are opened. Interlocks should always fail to safety.

2. In general, chassis and exposed metal parts should be earthed and bonded together. Conductors used for such purpose should be adequate for maximum anticipated fault current.
3. High voltage connectors should be of the type on which the live connector has recessed contacts so that accidental touching of the contacts is impossible.
4. If it is necessary to work on exposed live equipment, then it should always be under the direct supervision of the person working with it. Prominent warning notices stating that the equipment is live must be attached to the equipment.
5. Several types of valve, particularly of American manufacture, contain radioactive materials and care should be taken when disposing of them.
6. Selenium (in rectifiers) should be handled with caution because of its toxicity.
7. When a selenium rectifier burns out or arcs over, the area should be ventilated to remove fumes, particularly in the case of large rectifiers.
8. Rectifier tubes operating in circuits in which peak inverse voltages are around 16000 or over, produce X-rays, for which shielding should be provided.
9. The possibility of high-frequency burns is not confined to radio-frequency equipment, but can exist in large audio-frequency equipment because of high-frequency parasitic oscillations.
10. Before handling cathode-ray tubes, the high-voltage terminal to outer coating and earth should be shorted. The tubes should be carefully stored and carried to prevent breakage, and eye protection worn.
11. a) Only insulated or earthed shafts should protrude through chassis panels.

    b) As an added precaution short, well-recessed setscrews, preferably of nylon, should be used.
12. Whenever technical requirements permit, current-limiting resistors must be installed in series with the output of power supplies.
13. If power must be on while adjusting equipment, these precautions should be observed to minimize hazards:

    a) use insulated test prods;

    b) have another person who is cognisant of hazards and familiar with artificial respiration near you;

    c) stand on an insulating mat.
14. If it is necessary to disable an interlock, label it for the time it is inoperative.

STATIC ELECTRICITY

Liquids, gases, and solids can acquire static charges by their intimate contact and separation. Such charges are extremely hazardous and have accounted for serious damage, not only with organic liquids but also with dusts and powders (starch, sulphur) and compressed gases (including steam). The danger of spark discharge of any charged material is accentuated by the wearing of synthetic fibre clothing.

In liquid–solid systems where processes such as evaporation, agitation, pumping, and filtration are used, static charges can be generated and may accumulate on the containers. Tanks, pipelines, and any isolated conductors such as metal flanges should be earthed and bonded effectively. Not only bulk delivery wagons but portable drums and containers may accumulate a charge during the movement or transfer of their contents, so they should be adequately earthed. Flow rates should be restricted to a safe value and turbulence avoided by the use of submerged pipe exits.

Solids, dusts, and powders can accumulate static charges in operations such as grinding, sifting, blowing, extracting, rolling, etc. All equipment used for these operations should be bonded together and earthed.

Dispersion of Static

1. Bonding and earthing of equipment.
2. The use of liquids of low resistivity or the addition of resistivity reducing agents.
3. The use of constructional materials of low resistivity.
4. The control of pumping speeds to limit the rate of charging.
5. Humidification, or ionization of the atmosphere by radioactive emitters.
6. The use of submerged pipe exits where possible.

The above remarks are of a general nature and each case must be considered separately. Frequent surveys should be carried out with a Baldwin Dunlop Statigun, or similar instrument, to detect any accumulating charge.

CENTRIFUGES

The centrifuge rotor must be carefully balanced each time it is used. The centrifuge must be securely anchored and adequately shielded against accidental "flyaways". The top should be equipped with a safety switch which shuts off the power if it is inadvertently removed. The rotor compartment should be adequately ventilated if flammable materials are being used. The centrifuge should be placed so that accidental vibration cannot cause damage to other equipment. The rotor must never be stopped by hand and should always be kept clean as it can be seriously weakened by corrosion.

AUTOCLAVE AND PRESSURE EXPERIMENTS

General

These instructions are applicable to all types of pressure equipment. Any experiments involving vessels under pressure must be authorized by a senior member of staff and the Safety Officer must be consulted.

All types of laboratory pressure vessels and auxiliary equipment require careful selection, inspection, and testing. Equipment should be sited and precautions taken assuming that an explosion *will* occur. Switches and controls should be in a safe and preferably remote position. The maximum pressure attained should be calculated and untried reactions first tested on a small scale. Equipment must not be left under pressure without adequate supervision.

The construction materials of pressure vessels must be carefully chosen to withstand the reactants and conditions. Maximum working conditions of pressure and temperature must be determined for each vessel in use and should not be exceeded in service. The maximum safe working conditions must be clearly and indelibly marked on each vessel, together with a suitable identification mark for insurance purposes. The insurance company should be consulted in all cases to decide whether the equipment should be specially insured, in which case they will carry out the statutory testing.

Autoclaves

Autoclaves should be operated only in places intended for the purpose and instruments and controls be positioned behind a protective barrier.

A log book should be kept for each autoclave listing the following:

1. Pressure and temperature at which apparatus has been tested.
2. Maximum permissible working pressure and temperature.
3. Construction materials including lining.
4. Available volume.
5. Details of pressure relief equipment.
6. All subsequent information of periodic tests.
7. The nature of each reaction carried out, recorded briefly, as well as any faults that may have developed during use.

The test pressure must be at least 1.5 times the maximum admissible working pressure and testing should be repeated periodically. Special attention should be paid to any permanent deformation. Pressure gauges should not be allowed to record pressures greater than two-thirds of the scale range. Pressure relief equipment in the form of a bursting disc is generally more satisfactory than a spring-loaded valve. Each bursting disc or safety valve assembly must be connected to a separate discharge line which discharges to a safe place, preferably outside the building. A notice should be placed near the end of the discharge line to warn people to keep away from the vicinity.

Before using any autoclave the pressure gauge, valve, pressure relief equipment, and discharge lines must be carefully examined to ensure the absence of any obstruction.

For experimental work, the volume of the autoclave charge must not exceed one-half of the free space of the vessel. Both heat and pressure should be applied as gradually as possible and pressure should be released gradually. The tightening up of joints under pressure is potentially dangerous and should be avoided.

Before opening an autoclave the pressure inside must be reduced to atmospheric.

Glassware

The use of glassware for pressure reactions should generally be avoided unless absolutely necessary. When its use is unavoidable, the apparatus must be pressure tested with all safety precautions. Vessels should be thick walled and round bottomed and be screened with a stout wire mesh. The experiment should be conducted behind a protective shield which must give all-round protection to avoid endangering personnel or other equipment.

Due to the excessive danger of equipment failure, the general instructions for autoclaves should be followed. The maximum pressure attained should be calculated and apparatus designed and tested to a pressure of at least 1.5 times this value. Pressure relief equipment must be connected to a separate discharge line, which discharges to a safe place. Safety glasses and face shields should always be worn when working with glass pressure equipment.

FUEL GAS SUPPLY

The towns gas supply in the United Kingdom consists almost entirely of North Sea gas which contains approximately 93.5 per cent methane and 3.5 per cent ethane. Unlike coke-oven gas it is non-poisonous and odourless, and agents have to be added to the gas so that its presence can be detected by smell. It is of course highly flammable and both it and its combustion products can cause asphyxiation if the oxygen content is reduced to a low level. When used in admixture with compressed gases such as hydrogen, air, or oxygen, non-return valves must be fitted into the fuel gas line to prevent ingress of oxygen into the fuel line. This is a statutory requirement in the United Kingdom.

Repairs and alterations to the gas lines must only be carried out by qualified maintenance staff. In the case of gas rings, a metal tube should be used for the gas feed line to avoid ignition of rubber tube in close proximity to the burner. Flammable liquids are best heated in oil or water baths or by isomantle heaters.

HEATING BATHS

The electrical mantle type of heater has now largely superceded the water or oil bath heater. There is, however, still a demand for oil and water baths in the laboratory and desirable safety precautions will be considered. Oil baths present a very high fire risk and they should preferably not be allowed to run overnight or virtually unattended. An oil with a high flashpoint should be selected for use and typical of those that have been used are: Dow Corning silicone fluid DC200 or the range of polyalkylene glycols which have flash-points of around 255°C. Since oils deteriorate with heating, the flash-point should be checked at fairly frequent intervals. The electrical heating mechanism should be thermostatically controlled and have a safety cut-out device to prevent overheating.

COLD TRAP HAZARDS

If liquid nitrogen is the coolant, liquid air can condense in the cold trap. Two possible hazards may result from this: (a) a concentration of liquid oxygen may be formed which may explode violently if it comes into contact with organic or easily oxidizable materials; and (b) if the cold trap is isolated and allowed to warm up, it will explode. This situation may be avoided by pumping down the system before adding the liquid nitrogen and by excluding air leaks into the system. The appearance of a blue tint in liquid nitrogen is a direct indication of its contamination by liquid oxygen and it should be disposed of using the precautions generally used with liquid oxygen.

Liquid nitrogen and other liquefied gases should be handled carefully; at its extremely low temperature it can produce an effect on the skin similar to a burn. Eyes should be protected by a face shield or goggles and loose-fitting gloves be worn. Should any liquid nitrogen contact the skin or eyes, that part of the body should immediately be flooded with large quantities of unheated water and cold compresses applied. If the skin is blistered or the eyes have been affected, expert medical attention must be obtained.

Oxygen is removed from the air by liquid nitrogen stored in an open Dewar. Liquid nitrogen should only be stored and used in a well-ventilated place since the condensation of oxygen and evaporation of nitrogen can result in an oxygen deficient atmosphere. The oxygen content of the air must never fall below 16 per cent.

A common cooling mixture for cold traps is solid carbon dioxide (Drikold) and an organic solvent such as methanol or acetone. Too rapid an addition of solid carbon dioxide may cause spillage of solvent which is frequently highly flammable or toxic.

REFERENCES

1. Steere, N. V., ed. (1979). "Handbook of Laboratory Safety", 2nd edn., C.R.C. Press Inc., Boca Raton, Florida.
2. Spencer, E. W., Ingram, V. M., and Leventhal (1966). Electrophoresis: an accident and some precautions, *Science*, **152,** 1722.

FURTHER READING

"Effects of Current Passing through the Human Body", International Electrotechnical Commission, IEC Report, Publication 479, 1st edn. 1979. Bureau Central de la Commission Electrotechnique International, Geneva.

Ellis, J. G. and Riches, N. J. (1978). "Safety and Laboratory Practice", Macmillan Press, London.

Fawcett, H. H. and Wood, W. S., eds. (1965). "Safety and Accident Prevention in Chemical Operations", Interscience Publishers, London.

Grover, F. and Wallace, P. (1979). "Laboratory Organization and Management", Butterworths, London.

Handley, W., ed. (1977). "Industrial Safety Handbook", McGraw-Hill Book Company (UK) Ltd.

Hooper, E. (1980). "The Safe Use of Electricity", 2nd edn., The Royal Society for the Prevention of Accidents, Purley, Surrey.

"Mechanical Safety Devices for Laboratory Centrifuges", BS No. 4402 (1969) British Standards Institution, London.

Pieters, H. A. J. and Creyghton, J. W. (1975). "Safety in the Chemical Laboratory", 2nd edn., Butterworths, London.

"Recommended Safety Precautions for Handling Cryogenic Liquids" (1975). BOC Ltd. Available from Edwards High Vacuum, Manor Royal, Crawley, W. Sussex.

"Safety Precautions in the Use of Electrical Equipment" (1968). Imperial College of Science and Technology, London.

Stalzer, R. F., Martin, J. R. and Railing, W. E. (1967). Safety in the laboratory. *In* "Treatise on Analytical Chemistry", Part III, Vol. 1. (I. M. Kolthoff, P. J. Elving, and F. H. Stross, eds.) Interscience Publishers, New York.

"Static Electricity" (1963). Safety in Industry, Mechanical and Physical Hazards No. 8, Bulletin No. 256, US Department of Labor, Bureau of Labor Standards, Washington DC.

Steere, N. V. (1971). "Handbook of Laboratory Safety", C.R.C. Press Inc., Boca Raton, Florida.

Zabetakis, M. G. (1967). "Safety with Cryogenic Fluids", Heywood Books, London.

4

Laboratory Techniques

Before any work is started in the laboratory, a careful investigation must be made into the nature of reaction in order to determine whether a hazardous situation could develop. The investigation should start with the condition of the laboratory in which the work is to be performed. The floor should be clean, dry, and free from spillage of chemicals which could cause a person to slip. The gangways should be clear of obstructions, with particular attention being paid to escape routes. The bench should be clear of surplus equipment and adequate space be available to carry out the necessary work. This is a matter which is frequently overlooked with the result that equipment is crammed into a small space thereby increasing the chances of an accident. The services which are required for the work should be readily available and if special gases are required, the gas cylinders supplying them must be supported securely in cylinder stands. Electrical leads must be routed in a safe manner and held in position by clips. Safety equipment required during the work should be prepared beforehand and any warning devices must be tested. Personnel nearby who may be required to administer first aid or to carry out rescue work must be alerted and informed that their services may possibly be needed. The apparatus should be examined carefully to ensure that it is in good condition and capable of withstanding the temperatures and pressures reached during the course of the experiment. Broken or chipped glassware should be sent for repair or thrown away. In the case of work or experiments which have been previously performed, the hazards are usually known and the necessary precautions can be taken. Where new reactions are being investigated this is not so and it is advisable to work with the smallest possible amounts of material, taking all necessary precautions for personal protection. It is preferable to err well on the safe side rather than be guided solely by one's experience, as unpredictable results can frequently occur. The scale of the experiment can then be gradually increased until the required size is reached. Work of this type in which the outcome is unknown should not be carried out when alone in the laboratory.

Particular care must be taken when setting up large pieces of apparatus in the laboratory. The apparatus must be securely held using stands and clamps or, if particularly large, a special framework. The range of clamps available makes this a relatively straightforward job and adjustable laboratory jacks are a very useful method for supporting equipment. The apparatus should be assembled away from the edge of the bench and not in front of the service controls. These should be conveniently placed in such a position that they can be easily operated in the event of an incident. In the case of particularly large apparatus which extends well above head height, access via a platform may be necessary, which must be fitted with handrails extending to a height of 1 m from the base and with toeboards at least 5 cm in height.

Escape routes should be carefully planned and kept free from any obstruction. Stools should not be used when carrying out laboratory experiments where there is the slightest risk, since they may impede rapid escape or preventive action.

MACHINERY

Machinery, and laboratory equipment coming into the broad classification of machinery, normally consists of basic items such as belts on vacuum pumps rotating shafts, and must be guarded according to official regulations. Certain pieces of laboratory equipment are difficult to guard and in these cases it is recommended that a code of practice be developed to cover their operation.

LABORATORY STIRRERS

Laboratory stirrers can range from the simple type used to stir material in small beakers to large, high horsepower units used for milling or grinding purposes. The laboratory type of stirrer is generally of low power and is used for a variety of small-scale mixing operations. The flexibility of these stirrers is greatly reduced if they are safeguarded by the use of microswitches or by complete enclosure, and yet if used incorrectly they can cause accidents. In this case, it is advisable to develop a code of practice covering their operation, which all operators must read. Such a code of practice, developed after many years of experience, is shown in Table 4.1. This specifies the method for operating portable laboratory stirrers and, if all the precautions are taken, will ensure their operation is as safe as practicable in the circumstances of their use.

With regard to the heavier duty type of stirrer, two cases may be con-

Table 4.1. Code of practice—laboratory stirrers (portable).

1. All personnel must read, understand, and sign this "Code of practice" before being allowed to operate portable stirrers in the laboratories.
2. All procedures detailed below must be carried out each time a laboratory stirrer is operated.

Before Operation

3. Before setting up a portable stirrer ensure that it is disconnected from the socket.
4. Ensure that the blade (if separate) is properly fixed to the impeller shaft and tighten if necessary, taking care to avoid sharp edges on the impeller.
5. Ensure that the impeller shaft is properly fixed to the machine and tighten if necessary.
6. Ensure that the machine is standing evenly and is securely fastened to a suitable stand.
7. Ensure that the electrical cable is of a suitable length and in a position where it cannot be caught by moving parts or by the operator.
8. Ensure that there is adequate working space around the machine and unobstructed access to the bench.
9. Before starting the operation of the machine, ensure that
   a) the machine is switched off at socket,
   b) any switch mounted on machine is switched off,
   c) the variable speed switch on machine, or variac control is switched to zero.

Operation

1. Eye protection must be worn.
2. The machine *must not* be operated unless the blade of the impeller is immersed in liquid in the container.
3. Except where the rotating blade is completely protected the container must be clamped to the machine and not held by hand, and a check must be made that the impeller blade does not foul the container.

To Operate

i. switch on first at socket,
ii. then machine switch (if fitted),
iii. finally increase speed with variable switch or variac control.

In case of any difficulty, switch off immediately at the most convenient safe point and at the mains socket, then remove plug.

If container becomes unclamped, do not under any circumstances attempt to hold it. Stop the machine immediately.

At the end of mixing, switch off stirrer and remove plug before raising impeller.

Switch off at all possible positions and clean up.

Precautions

1. Do not allow solvent based mixtures to overheat.
2. Ensure that the stirrer motor is powerful enough to carry out the operation without overheating.

sidered. The high-speed powerful stirrer introduces a noise problem as well as a rotational problem and in these cases a sound-proofed cupboard is recommended. Such a cupboard is shown in Fig. 4.1. The stirrer is of Bosch manufacture, operating at the high speed of rotation of 12 000 rev/min. The arrangements for securely clamping the stirrer and beaker are shown in the right-hand cupboard, the left-hand cupboard door being closed as the stirrer is in operation. The apparatus is set up and the cupboard closed. The operating controls are outside the cupboard and they incorporate a time switch for stopping the stirrer at a specified time. A microswitch can be fitted to the door so that the stirrer motor is disconnected when the door is opened. The switch should be in the correct failsafe mode as recommended by the manufacturer.

In some cases, it is necessary to have access to the vessel while the stirrer is in operation. Such a heavy duty commercially available stirrer is shown in

Fig. 4.1. High-speed stirrer protection. The stirrers are enclosed in cupboards lined with sound-proofing, and with the containers rigidly mounted; the timer and isolation controls are mounted in the front of the cupboards. If required, microswitches can be fitted to the doors so that the stirrers are prevented from being operated when the doors are open.

Fig. 4.2. This apparatus is used in paint manufacture, where pigments are added during the stirring operation. The stirrer blades have sharp cutting edges and two microswitches have been added so that the motor cannot be started unless there is a vessel placed in the clamp and the stirrer and motor are lowered into position. Furthermore, as it has been known for the stirrer blades to slice through an incorrectly positioned vessel, the two clamps are of thick gauge metal to prevent the blades being exposed. It was noted also that there was an unprotected spindle above the motor in the apparatus as received and this has been protected by a guard. These modifications are typical of those that frequently have to be made to bring commercially available apparatus up to an acceptable standard of safety. However, condi-

Fig. 4.2. A heavy duty laboratory stirrer. Microswitches are fitted to: (a) the container clamp so that the motor cannot be started unless the container is in position and, (b) the column assembly so that it must be lowered into position before the motor can be started.

tions in the UK since the introduction of the Health & Safety at Work Act have vastly improved and manufacturers are becoming increasingly aware of their duties as laid down in the Act.

It is acknowledged that certain pieces of apparatus present difficulties in that if they are completely protected against the possibility of accident they cannot be used. It is necessary in these cases to instal as many safety precautions as possible, to ensure that detailed procedures are laid down for safe working, that operation is restricted to personnel who are fully conversant with the apparatus, and that suitable personal protection is worn.

ROLL MILLS

Roll mills which are used, for example, for incorporating pigment into thermoplastic resins are a form of laboratory machinery which requires careful consideration, bearing in mind that access to the rolls is necessary during the course of an experiment. Mills of this type should be provided with a number of stop switches of the press-button type positioned round the machine so that they can be operated by the knee or foot and not with the hand. Additional switches should be placed immediately over the "nip" of the rolls; when these are pressed the rolls stop immediately and then go into reverse, so that anything caught is released. A typical code of practice covering the operation of triple roll mills is shown in Table 4.2.

Table 4.2. Code of practice for operating triple roll mills.

Ensure that

1. Floor around machine is clear and dry.
2. All floppy clothing, e.g. ties and long hair, is suitably secured to prevent dangling near the rolls.
3. Oil reservoirs are at least one-third full.

Operation 1—Cleaning the Machine

In order to avoid rusting of the rollers it is necessary to grease them when not in use. This grease must be removed before the machine can be operated.

1. Remove side limiters and safety bar.
2. Remove scraper—taking great care as the blade is extremely sharp.
3. Open back roll and front roll from centre roll to give a gap of at least 0.5 in between each roll.
4. Switch oil to drip feed.
5. Switch cooling water on to slow steady flow.
6. Start machine and clean rolls with white spirit and paper tissues, keeping the fingers well clear of the inward running "nip".
7. STOP THE MACHINE.

8. Clean side limiters and safety bar with white spirit and paper tissues—these may have been inadvertently contaminated with grease from rollers.
9. Clean scraper with white spirit and Ffitch brush.
   DO NOT CLEAN BLADE BY HAND.

Operation 2—Operating the Machine

1. Close back and front rolls to centre roll to give a just visible gap between rolls ensuring that the gap is even along the length of the rolls.
2. Replace side limiters and safety bar, adjust to required width between limiters and tighten.
3. Replace scraper and adjust to just contact front roller.
4. Place part of millbase paste on to the back roll to ensure immediate coating of rolls when started.
5. START MACHINE.
6. Place remainder of millbase paste on to the back roll and on top of the bar by means of a spatula or wallpaper scraper.
7. Millbase must not be pushed between the rolls with spatula or scraper.
8. Machine must not be operated without side limiters and safety bar in position.
9. Adjust gap between rear roll and centre roll to give thin even band of paste on centre roll.
10. Adjust gap between front roll and centre roll to give thin even band of paste on front roll.
11. Adjust scraper to remove paste evenly.
12. Collect milled paste in suitable container or retain on scraper until returned for further pass.
13. STOP THE MACHINE.
14. If more than one pass is required to give the necessary dispersion repeat operations 4, 5, 6, 12, and 13.
15. *N.B.* Some pastes may more conveniently be returned to the back roll without stopping the mill in complete safety, but at no time should the mill be run with dry rolls.

Operation 3—Cleaning the Machine

1. Remove scraper—clean with appropriate solvent and Ffitch brush.
   DO NOT CLEAN BY HAND.
2. Remove side limiters and safety bar—clean with appropriate solvent and paper tissues.
3. Open the rolls to give at least a 0.5 in gap between each roll.
4. Start machine and clean rolls with appropriate solvent and tissues ensuring fingers are kept well clear of inward running "nip".
5. STOP MACHINE WHEN CLEAN.
6. START MACHINE and grease with petroleum jelly keeping fingers well clear of "nip".
7. STOP MACHINE.
8. Switch off oil feed to bearings.
9. Switch off cooling water.
10. Replace side limiters, safety bar, and scraper.

All personnel intending to use this machine must report to the laboratory steward in the mill room before use. They must read the instructions and abide by them.

SCREENS

If there is the slightest risk of being injured during the course of an experiment, then a protective screen must be used. This should be of such a size to protect adequately the person performing the experiment; in addition, passers-by and persons working at the other side of the bench must be protected. Screens should be made from shatter-proof laminated glass or 6 mm polycarbonate plastic, and held securely in a stout metal or wooden frame. It should be remembered that they only afford protection as long as the screen is between the whole of the body and the apparatus. To facilitate this it is advisable to have all the controls for the apparatus in front of the screen. If, however, it is necessary to place hands or arms behind the screen then these should be carefully protected by gloves of suitable length. One operation for which screens must always be used is vacuum distillation. A personal experience illustrates a situation in which a screen prevented serious injury. A vacuum distillation was being observed in which the boiler was a 2-litre round-bottom flask and the distillation had nearly reached its end. There was suddenly a loud bang and the flask disappeared. Small pieces were found all over the laboratory, underlining the need for all round protection and demonstrating the fact that implosions can be as hazardous as explosions.

COMMINUTION OF MATERIALS

Laboratories, particularly those involved in the examination and analysis of materials, may possess a range of devices for grinding or milling samples. These can include face grinders, jaw mill crushers, ball mills, mechanical mortars and pestles, and hammer mills, all of which are designed to break up materials and are equally efficient at crushing parts of the human body. Each piece of equipment must be carefully studied to ensure that maximum protection is afforded to the operator. In extreme cases it may be necessary to enclose the whole apparatus in a box protected by microswitches so that the equipment only operates when the box is closed. Similar remarks apply to mixers. The stirring type has already been considered but hazards also exist with those that consist of an eccentric rotating vessel. Here the main danger is associated with parts of the body being trapped between the vessel and the wall or bench, and they should be isolated from such hazards by a wire mesh screen fitted with a microswitch so that removal of the screen automatically stops the mixer. Generally the safeguarding of drive mechanisms, belts, rotating shafts, etc., has been carried out by the equipment manufacturer but it is well worth looking closely at these items to ensure that all hazards have been eliminated.

CORRECT USE OF APPARATUS

In procedures which have already been laid down and used many times, it may safely be assumed that the specification for the design and type of apparatus is satisfactory. In these cases it is only necessary therefore to conform rigidly to the procedural details laid down for the experiments. In research work, where novel procedures are being attempted, care has to be taken to specify exactly the type of apparatus needed to ensure that the work is carried out safely. The researcher must use his judgement when designing apparatus and must call on expert help if unsure of any aspect of the work. When the equipment is obtained from the manufacturer it must be carefully checked to ensure that it is up to specification and that it will fulfil the duty for which it was selected. When the whole apparatus is assembled he must carry out further checks so that he is confident that no mishap will occur. Every effort must be made to ensure that no unexpected event can occur; if there is a possibility that a hazardous situation may arise and it cannot be allowed for in the design of the equipment then measures must be taken to protect the operators. A typical example of the use of measures of this type is the employment of bursting discs or pressure relief valves in pressure experiments where care should be taken to ensure that they operate well below the design limitations of the apparatus and discharge to a safe position.

Equipment can deteriorate with use and should be examined carefully for possible flaws beforehand. This applies particularly to glassware, especially if it is to be used under vacuum or for pressure work.

In vacuum work the vacuum source should be protected by a trap system to avoid contamination of the oil in the vacuum pump or, if a water pump is used as the source, to avoid water sucking back into the experiment. In pressure experiments in which gas cylinders are used as the source of pressure, a pressure relief device such as a lute, bursting disc, or pressure relief valve should be used in case the gas cylinder regulator ceases to function.

There is a range of heating equipment now available and care should be taken to ensure that the correct form is used. There is still a use for the Bunsen burner in laboratories where the risk of fire is small as they provide a rapid and easily controllable source of heat. Isomantles provide a good all-round source of heat and have become increasingly popular during the last few years. Controlled temperature hot plates are particularly useful for heating PTFE equipment which can be safely used up to a temperature of 250°C.

HANDLING TOXIC AND REACTIVE CHEMICALS

The title of this section is rather a misnomer as chemicals should never be handled, if at all possible. All chemicals have some degree of risk attached to their use and an important criterion is the amount of the chemical required to cause death which can be as little as 50 mg of hydrocyanic acid or 250 mg of cyanides. The first step to take when working with chemicals is to find out as much information as possible about them from published sources. A fairly comprehensive list of these sources is given at the end of the chapter; if the information sought is not included in these then the manufacturer, who, in the UK, now has a legal obligation to inform users of his materials of potential hazards, should be approached. In the case of new materials which have been synthesized by the research worker such information is not available and judgement must be exercised in predicting whether a material will be hazardous. This may not always be predictable and the compounds may have to be subjected to exhaustive testing; this is always the case if the compound is to be marketed and to carry out full-scale toxicological testing is a very expensive task. The extreme care in working with hazardous chemicals is exemplified in radiochemical and certain toxicologic laboratories. It is impossible to deal with both these types of laboratory in this book and the reader is referred to one or other of the books listed in the bibliography at the end of the chapter for full details. Suffice to say that the laboratory is isolated completely from the outside world by a door interlock system and workers must change their clothes before entering the laboratory. Sophisticated remote handling equipment is used to manipulate the equipment and all gaseous, liquid, and solid effluents are carefully monitored before they are discharged to atmosphere. Some effluents may have to be destroyed before they can be disposed of and certain radioactive effluents may have to be encased and dumped at sea.

In working with hazardous chemicals the first requirement is to know exactly the nature of the hazard. All personal precautions such as the wearing of gloves, eye protection, and special clothing should be taken. Any antidotes should be prepared and made readily available for instant use. Flammable materials should be kept to the minimum quantity, sources of ignition excluded and the correct type of fire extinguisher placed nearby ready for use. Materials with a flashpoint below ambient temperature require very careful consideration. The vapour pressure/temperature curve should be carefully examined to check whether an explosive concentration of the substance in air can develop. The lower explosive limit is in general 1 per cent for most flammable materials in air; in oxygen-enriched atmospheres, however, it may be less. If a concentration of this order is likely to be achieved then the ventilation must be increased. Tests should also be carried out using one of the many instruments on the market to check the flammability of the atmosphere and if this approaches 20 per cent of the lower explosive limit, action should be taken. It is rare for the upper explosive limit to be exceeded but it is equally important to act if this occurs. The possibility of a flash-back from an ignition source remote to the experiment has already been mentioned and all possible sources of ignition should be removed.

Highly reactive or explosive materials require very careful handling and reactions involving them should be kept to the smallest possible size. Several reactions can be carried out if a larger amount of material is required. Precautions should be taken on the assumption than an explosion will occur. Highly exothermic reactions should be classified in this category, particular care being taken to ensure that cooling systems do not fail with the result that reactions get out of control. Disposal of waste from such reactions requires special precautions as reactive compounds can be generated by side reactions. The literature is liberally sprinkled with descriptions of accidents of this type, perhaps the most noteworthy being the formation of the highly explosive silver nitride in some photographic solutions under certain conditions and in silvering solutions.

With regard to toxic chemicals, the question must be asked whether a safer substitute can be used. An example of this is the substitution of the carcinogenic O-tolidene by p-amino-N:N-diethylaniline sulphate in the test for traces of chlorine. Personal cleanliness reduces the possibility of ingestion and the wearing of protective clothing eliminates absorption of the compound through the skin. Protective clothing should always be examined carefully before use as, for example, minute pinholes in gloves can cause serious injury. The danger may be virtually completely eliminated by carrying out the work in an efficient fume cupboard; in this case, however, care should be taken to ensure that the exhausted material from the fume cupboard does not cause problems elsewhere. The literature on toxicity of chemicals is vast and this should be consulted. A fairly comprehensive list is given at the end of the chapter.

Finally care should be taken when handling substances which may react violently when mixed. This should also be borne in mind when storing chemicals and a list of such compounds is given in Table 4.3, with a further list of chemicals which on reaction produce a third toxic compound given in Table 4.4.

Table 4.3. Partial list of incompatible chemicals (reactive hazards).

Substances in the left-hand column should be stored and handled so they cannot possibly accidentally contact corresponding substances in the right-hand column under uncontrolled conditions, when violent reactions may occur.

Acetic acid	Chromic acid, nitric acid, peroxides and permanganates.
Acetic anhydride	Hydroxyl-containing compounds, ethylene glycol, perchloric acid.
Acetone	Concentrated nitric and sulphuric acid mixtures.
Acetylene	Chlorine, bromine, copper, silver, fluorine and mercury.
Alkali and alkaline earth metals, such as sodium, potassium, lithium, magnesium, calcium, powdered aluminium	Carbon dioxide, carbon tetrachloride and other chlorinated hydrocarbons. (Also prohibit water, foam, and dry chemical on fires involving these metals—dry sand should be available.)
Ammonia (anhyd.)	Mercury, chlorine, calcium hypochlorite, iodine, bromine and hydrogen fluoride.
Ammonium nitrate	Acids, metal powders, flammable liquids, chlorates, nitrites, sulphur, finely divided organics or combustibles.
Aniline	Nitric acid, hydrogen peroxide.
Bromine	Ammonia, acetylene, butadiene, butane and other petroleum gases, sodium carbide, turpentine, benzene and finely divided metals.
Calcium oxide	Water.
Carbon, activated	Calcium hypochlorite, other oxidants.
Chlorates	Ammonium salts, acids, metal powders, sulphur, finely divided organics or combustibles.
Chromic acid and chromium trioxide	Acetic acid, naphthalene, camphor, glycerol, turpentine, alcohol and other flammable liquids.
Chlorine	Ammonia, acetylene, butadiene, butane and other petroleum gases, hydrogen, sodium carbide, turpentine, benzene and finely divided metals.
Chlorine dioxide	Ammonia, methane, phosphine, and hydrogen sulphide.
Copper	Acetylene, hydrogen peroxide.
Fluorine	Isolate from everything.
Hydrazine	Hydrogen peroxide, nitric acid, any other oxidant.

Table 4.3 *contd*

Hydrocarbons (benzene, butane, propane, gasoline, turpentine, etc.)	Fluorine, chlorine, bromine, chromic acid, peroxides.
Hydrocyanic acid	Nitric acid, alkalis.
Hydrofluoric acid, anhyd. (hydrogen fluoride)	Ammonia, aqueous or anhydrous.
Hydrogen peroxide	Copper, chromium, iron, most metals or their salts, any flammable liquid, combustible materials, aniline, nitromethane.
Hydrogen sulphide	Fuming nitric acid, oxidizing gases.
Iodine	Acetylene, ammonia (anhyd. or aqueous).
Mercury	Acetylene, fulminic acid,* ammonia.
Nitric acid (conc.)	Acetic acid, acetone, alcohol, aniline, chromic acid, hydrocyanic acid, hydrogen sulphide, flammable liquids, flammable gases, and nitratable substances.
Nitroparaffins	Inorganic bases, amines.
Oxalic acid	Silver, mercury.
Oxygen	Oils, grease, hydrogen, flammable liquids, solids, or gases.
Perchloric acid	Acetic anhydride, bismuth and its alloys, alcohol, paper, wood, grease, oils.
Peroxides, organic	Acids (organic or mineral), avoid friction, store cold.
Phosphorus (white)	Air, oxygen.
Potassium chlorate	Acids (see also chlorates).
Potassium perchlorate	Acids (see also perchloric acid).
Potassium permanganate	Glycerol, ethylene glycol, benzaldehyde, sulphuric acid.
Silver	Acetylene, oxalic acid, tartaric acid, fulminic acid, ammonium compounds.
Sodium	See alkali metals (above).
Sodium nitrite	Ammonium nitrate and other ammonium salts.
Sodium peroxide	Any oxidizable substance, such as ethanol, methanol, glacial acetic acid, acetic anhydride, benzaldehyde, carbon disulphide, glycerol, ethylene glycol, ethyl acetate, methyl acetate and furfural.
Sulphuric acid	Chlorates, perchlorates, permanganates.

*Produced in nitric acid–ethanol mixtures.

Table 4.4. Partial list of incompatible chemicals (toxic hazards).

Substances in the left-hand column should be stored and handled so that they cannot possibly accidentally contact corresponding substances in the centre column, because toxic materials (right-hand column) would be produced.

Arsenical materials	Any reducing agent	Arsine*
Azides	Acids	Hydrogen azide
Cyanides	Acids	Hydrogen cyanide
Hypochlorites	Acids	Chlorine or hypochlorous acid
Nitrates	Sulphuric acid	Nitrogen dioxide
Nitric acid	Copper, brass, any heavy metals	Nitrogen dioxide (nitrous fumes)
Nitrites	Acids	Nitrous fumes
Phosphorus	Caustic alkalis or reducing agents	Phosphine
Selenides	Reducing agents	Hydrogen selenide
Sulphides	Acides	Hydrogen sulphide
Tellurides	Reducing agents	Hydrogen telluride

*Arsine has been produced by putting an arsenical alloy into a wet galvanized bucket. (Tables 4.3 and 4.4 are reprinted from "Hazards in the Chemical Laboratory", see Bibliography.)

FURTHER READING

Brenner, M. (1976). Hazards of school science, *New Sci.* **69,** 550–2.

"Code of Practice for Chemical Laboratories" (1976). Royal Institute of Chemistry, London.

Green, M. E. and Turk, A. (1978). "Safety in Working with Chemicals", Macmillan Publishing Co. Inc., New York and Collier Macmillan Publishers, London.

"Laboratory Safety Handbook" (1976). Sanderson Chemical Consultants Ltd, Thornaby, Cleveland, UK.

"Laboratory Safety Management" (1979). Chemical and Allied Products Industry Training Board, Staines, UK.

"Safety in Chemical Laboratories and in the Use of Chemicals" (1970). Imperial College of Science and Technology, London.

"Safety in Laboratories" (1974). Ciba-Geigu (UK) Ltd, 30 Buckingham Gate, London, SE1E 6LH.

"Safety Manual" (1973). The University of Manchester Institute of Science and Technology, Manchester.

"Safety in Science Laboratories" (1978), 3rd edn., HMSO, London.

Schuerch, C. (1972). Safe practice in the chemical laboratory, **49,** Ay683-A685, 637–A639.

Steere, N. V. (1971). "Handbook of Laboratory Safety", CRC Press Inc., Boca Raton, Florida, USA.

Taylor, P. J. (1978). Developing an OSHA acceptable academic chemistry department, *J. Chem. Educ.* **55,** Part 12, A439–A441.

Berman, E. (1980). "Toxic Metals and their Analysis", Hayden, London, Philadelphia, Rheine.

Wawzonek, S. (1978). Safety practice in the undergraduate organic chemistry laboratory, *J. Chem. Educ.* **55,** Part 2, A71–A74.

Young, J. R. (1971). Responsibility for a safe high school laboratory, **48,** A349–A356.

A SELECTED BIBLIOGRAPHY ON HAZARDOUS CHEMICAL SUBSTANCES

"Advances in Modern Technology" (1977). Goyer, R. A. and Mehlman, M. A., eds. Vol. 2, Toxicology of Trace Elements, John Wiley & Sons, New York and London.

American Conference of Governmental Industrial Hygienists, Documentation of threshold limit values for substances in the workroom air, 1979.

"Auto-ignition Temperatures of Organic Chemicals", Hilado, C. J. and Clark, S. W., Chemical Engineering (1972), 75–80, Sept. 4th.

Carcinogens, Suspected, A subfile of the National Institute for Occupation Safety and Health (NIOSH) toxic substances list, US Department of Health, Education and Welfare.

"The Care, Handling and Disposal of Dangerous Chemicals", 2nd ed. (1970). Garton, P. J., Publishers (Aberdeen) Ltd.

Chemical Safety Data Sheets, Manufacturing Chemists Association Inc., Washington DC.

"The Chemistry of Industrial Toxicology", 2nd ed. (1959). Elkins, H. B., J. Wiley & Sons, New York and London.

Chemical Toxicology of Commercial Products, 4th ed. (1976), Gosselin, R., *et al.* Williams & Wilkins Co.

Codes of Practice for Chemicals with Major Hazards, Chemical Industry Safety & Health Council of the Chemical Industries Association.

Compilation of Odour Threshold Values in Air and Water (1977). Germert, L. J. Van, Nettenbreijer, A.H., National Institute for Water Supply, Voorburg; Central institute for nutrition and food research TNO Zeist, The Netherlands.

"Condensed Chemical Dictionary", 9th ed. (1977). Hawley, G. A., Van Nostrand Reinhold, New York.

"Dangerous Properties of Industrial Materials", 5th ed. (1979). SAX.N.1, Van Nostrand, New York.

"Diseases of Occupation", 6th ed. (1978). Hunter, D., Hodder and Stoughton, London.

"Effects of Exposure to Toxic Gases", Baker & Mossman, Matheson Gas Products Inc., USA.
"An Encyclopaedia of Chemicals and Drugs", The Merck Index, 9th ed. (1976). Merck & Co.
"Environmental and Industrial Health Hazards, A practical guide" (1976). Trevithick, R. S., William Heinemann Medical Books, London.
"The Extra Pharmacopeia", 27th ed. (1977). Martindale, Pharmaceutical Press.
Flammability characteristics of combustible gases and vapours (1964). Zabetakis, M. G., US Bureau of Mines, Bulletin 627, Washington DC.
"Handbook of Chemistry and Physics", ed. Weart, R. C., 61st ed. (1980). The Chemical Rubber Publishing Co, New York.
"Handbook of Industrial Toxicology" (1976). Plunkett, E. R., Hayden, New York.
"Handbook of Laboratory Safety", 2nd ed. (1970). Steere, N. V., The Chemical Rubber Publishing Co., New York.
"Handbook of Poisoning: Diagnosis and Treatment" (1971). Dreisbach, R. H., Lange Medical Publications, Los Altos, California.
"Handbook of Reactive Chemical Hazards" (1975). Bretherick, L., Butterworths, London.
"Handling Chemicals Safely", 2nd ed. (1980). Published by the Dutch Association of Safety Experts, the Dutch Chemical Industry Association and the Dutch Safety Institute.
"Hazards in the Chemical Laboratory", 3rd ed. (1981). Bretherick, L., The Royal Society of Chemistry, London.
"Industrial Hygiene and Toxicology", Vol. II, 2nd ed. (1963). Palty, F. A., Interscience Publishers, New York.
"Industrial Health Technology" (1958). Harvey, B. and Murray, R., Butterworths, London.
International Labour Office Encyclopaedia of Occupational Health and Safety, 2 Vol. (1971). International Labour Office, Geneva.
Laboratory Waste Disposal (1969). Manufacturing Chemists Association, Washington DC.
National Institute for Occupational Safety & Health (NIOSCH) Registry of Toxic Effects of Chemical Substances, Annual Editions, US Department of Health, Education and Welfare.
Recommendations of the International Commission on Radiological Protection (1965). Pergamon Press, Oxford.
"Solvents Guide", 2nd ed. (1963). Marsden, C. and Mann, S., Cleaver Hume Press.
Threshold Limit Values for 1979, Guidance Note EH15/79, The Health and Safety Executive, HMSO, London.
Toxic and Hazardous Industrial Chemical Safety Manual (1976). The International Industrial Technical Information Institute, Tokyo.
"Toxicity of Industrial Organic Solvents" (1952). Browning, E., Edward Arnold, London.

"Toxicity of Drugs and Chemicals" (1973). Deichmann, W. B. and Gerard, H. W., Academic Press, London and New York.

"Toxicity of Industrial Metals", 2nd ed. (1969). Browning, E., Butterworths, London.

"Toxicity and Metabolism of Industrial Solvents" (1965). Browning, E., Elsevier, Amsterdam.

Treatment of Common Acute Poisonings, 4th ed. (1979). Matthew, H. and Lawson, A. A. H., Churchill Livingstone, London.

5

Classification of Commonly Used Chemicals and their Hazards

There is a good deal of truth in the claim that all chemicals are hazardous. Two simple chemicals which are present in the earth's atmosphere and crust provide good examples of the validity of this statement. Oxygen is a gas which is essential to life but in an atmosphere of pure oxygen, combustion takes place with incredible speed. An example of this occurred in one of the American spacecraft missions when virtually pure oxygen was used in the capsule. A slight electrical spark immediately caused everything combustible in the capsule to catch fire and burn vigorously, with the result that all three astronauts perished. The system was later modified and a reduced concentration of oxygen circulated.

Water is the second compound to be considered which for a non-swimmer out of his depth can produce a very hazardous situation that may result in his or her death. Severe stomach cramp can be produced if a person deprived of water for some time drinks large amounts in an uncontrolled manner. It is important, therefore, to consider not only the nature of the chemical but also the amount taken. Some drugs, for example, are effective in curing a person up to a certain dose rate but if this is exceeded then harmful side effects result. Compounds of mercury, arsenic, and antimony which to the layman are considered to be highly poisonous have been used in the treatment of ailments. It is therefore necessary to quantify the hazardous properties of chemicals so that correct precautions can be taken to ensure their safe use.

It is difficult to imagine, with present-day stringent controls on the use of chemicals, what the chemical factory looked like in the eighteenth and nineteenth centuries (Fig. 5.1). Corrosive and toxic fumes were allowed to pour out from low-level chimneys and waste by-products were dumped on land round the factory, with the result that all vegetation and wildlife over a considerable area were killed. Life expectancy for the chemical worker in those days must have been very short. The indiscriminate use of chemicals in other industries produced much pain and suffering and many examples

Fig. 5.1. A view of chemical factories and adjacent houses in Widnes taken towards the end of the nineteenth century.

could be quoted of diseases which are fortunately only a rarity these days, if indeed they exist at all. Compare this situation with that in the modern chemical factory with its strict controls on environmental pollution and the sophisticated methods for measuring concentrations of hazardous chemicals and it will be evident that remarkable progress has been made.

The hazards associated with chemicals can be classified under the general headings: toxicity, flammability, explosibility, and those which may be considered under the heading "generally offensive". The latter includes chemicals which have an unpleasant odour, or produce smoke, dust clouds, etc.

CLASSIFICATION OF TOXICITY

Threshold Limit Value

As mentioned in the opening paragraph, the hazardous properties of chemicals are largely affected by the size of the dose and it is necessary, therefore, to have systems for measuring the degree of their toxicity. The concept of threshold limit value (TLV) has been introduced to give an indication of the concentration of a chemical substance in the atmosphere which can be considered as being without hazard in a person's normal working life. The TLV of a substance is quoted in two concentration units as (a) parts of vapour or gas per million parts of contaminated air by volume at 25°C and 760 mm mercury pressure, or (b) approximate milligrams of substance per cubic metre of air. TLV-TWA (threshold limit value—time weighted average) is defined as being the time weighted average concentration for a normal 8-hour working day or 40-hour working week to which nearly all workers may be repeatedly exposed, day after day, without adverse effect. The qualification in this definition, "nearly all workers", should be noted. Some persons may experience irritation at lower concentrations than the TLV-TWA in which case additional precautions will need to be taken to reduce the concentration of the irritant to a tolerable level. The TLV-TWA should not be viewed as a sharp dividing line between safe and unsafe conditions, but more as a recommended concentration which should not be exceeded and, if it is possible to work at a considerably lower level without undue trouble or expense then this should be done. Time weighted averages permit excursions in concentration above the limit, provided that they are compensated by equivalent periods below the limit. The exact extent of the increase in concentration cannot be generally defined as it depends on the particular compound, each of which must be given separate consideration.

In 1976 a further concept was introduced, namely the Short Term Exposure Limit (STEL) value. This is defined as the maximal concentration to which workers can be exposed for a period up to 15 min continuously without suffering from:

(a) irritations;
(b) chronic or irreversible tissue change; or
(c) narcosis of sufficient degree to increase accident proneness, impair self rescue, or materially reduce work efficiency.

No more than 4 periods up to this concentration per day are permitted with at least 60 min at lower concentration between each. The daily TLV-TWA must not be exceeded.

For substances with a high hazard rating, the concept of Threshold Limit Value-Ceiling (TLV-C) has been introduced and these concentrations must not be exceeded, even instantaneously. It is, in fact, important to note that if either of these three designated concentrations, TLV-TWA, TLV-STEL, or TLV-C, is exceeded then a potentially hazardous situation may exist. Data for the three types of TLV are given in a list of approximately 700 compounds, published annually by the American Conference of Governmental Industrial Hygienists (ACIGH) and reproduced, with modifications according to UK practice, as Guidance Notes by the Health and Safety Executive in the UK. The information is based on industrial experience, human, and animal studies, and is subject to annual revision. The data have been drawn up with the average person in mind and it is inevitable some will have a greater or lesser sensitivity than others. These factors should be borne in mind when assessing the degree of hazard of a particular environment. TLV data are published to provide information about possible hazards and should be accepted as guidelines for good practice when considering atmospheric pollution. The possibility of variation in sensitivity of different persons should be considered; in those instances where sensitivity is enhanced, detection of the contaminant is facilitated, so the problem is to reduce the concentration and to protect the individual by the use of respirators, etc. For persons with reduced sensitivity, there is a danger that they may be in a hazardous atmosphere without knowing it and continuous monitoring for the substance is then essential.

TLV data for a few chemicals are listed in Table 5.1. These have been selected to illustrate the differences between the three different types. Appendices to the ACIGH contains information about carcinogens, substances of variable composition, mixtures, and the calculation of their TLV, resistance particulates, and asphyxiants.

Table 5.1. TLV data selected to show the variation that can exist between TWA and STEL values.

	TWA Adopted Values		*STEL Tentative Values*	
	ppm	mg M^{-3}	ppm	mg M^{-3}
Bromine	0.1	0.7	0.3	2
Chlorine	1	3	3	9
Cresol, all isomers—skin	5	22	—	—
Dimethyl aniline—skin	5	25	10	50
Ethanolamine	3	8	6	15
Methyl cyclo hexanol	50	235	75	350
Sulphur hexafluoride	1000	6000	1250	7500
C. formaldehyde	2	3	—	—
C. phenylphosphine	0.05	0.25	—	—

N.B. The designation "skin" refers to the potential contribution to the overall exposure by the cutaneous route including mucous membranes and eye, either by airborne or direct contact with the substance. Note that vehicles can alter skin absorption. The designation is intended to suggest appropriate measures for the prevention of cutaneous absorption so that the threshold limit is not invalidated. C = TLV-C ratings. (Reproduced by permission of the American Conference of Governmental Hygienists Inc.)

Dose Rating

The foregoing section is concerned with defining limits of concentration of compounds in the atmosphere to ensure that hazards associated with them are reduced. When the inspection of chemicals is considered, it is desirable to have a standard notation so that their toxicity can be directly compared. The most commonly used notation is the median lethal dose or LD^{50}. This is the dose which when administered to laboratory animals kills half of them, and it is expressed as milligram of substance administered per kilogram of animal weight. In comparing the relative effects of substances care must be taken to ensure that all the conditions such as type of animal, method of dosing, duration of treatment, are the same otherwise comparisons cannot be made. Extrapolation of animal results to human beings must be made with caution and the LD_{50} results should be taken to give only a rough indication of the comparable degrees of toxicity. For example, if a chemical is as exactly toxic to man as to rats (this is clearly not always true) and the LD_{50} for rats is 20 then the LD_{50} for a 70 kg man will be 1.4 g. This is only an approximation and should be taken to be a rough guide to the toxicity to man of the chemicals. So far death of the animal has been taken as the end result of the test; other factors may be considered which can be measured on a scale of increasing severity, e.g. onset of blindness, loss of faculties, anaemia, and so on; in this case, the dose to produce the effect if 50 per cent of the animals is called the ED_{50} or effective dose.

METHODS OF ABSORPTION

Apart from deliberate injection through the skin, there are three main routes by which individuals can be exposed to chemicals: through the skin, orally, and by inhalation.

Through the Skin

It is hard to convince the layman that chemicals can pass through unbroken skin and be absorbed into the bloodstream, but this can take place quite readily with a number of chemicals. There are certain parts of the skin which are more active than others, namely sweat and sebaceous glands, hair follicles, etc. and any area against which clothing rubs is particularly vulnerable. Similarly, confinement of the chemical in contact with the skin, e.g. if it is trapped in a boot or shoe, increases the chances of injury. Localized irritation of the skin is the commonest form of complaint and is enhanced by chemicals which absorb moisture and dry the skin. Corrosive chemicals cause varying degrees of injury, from relatively mild attack to the severe burns caused by hydrofluoric acid. Toxic chemicals may be absorbed into the bloodstream after passage through the skin and mercury is typical of chemicals which behave in this way. The authors recall a particularly severe case of mercury poisoning which occurred many years ago. Suck-back had occurred in a vacuum system consisting of mercury diffusion pumps backed by rotary oil pumps with the result that an emulsion of oil and mercury was present in the rotary oil pumps. A fitter dismantled, cleaned, and reassembled the pumps over a period of several days using no hand or arm protection. Several months later he developed the classic symptoms of mercury poisoning, probably due to absorption of mercury through the skin, facilitated by the oil.

The eye is a particularly sensitive area of the body and is irritated by the physical pressure of even the smallest object. When the material is corrosive or toxic the pain and extent of the injury are magnified greatly. Impairment or loss of vision is a serious consequence and an excellent reason for the compulsory wearing of eye protection in laboratories.

Inhalation

Of all the methods by which toxic materials can be absorbed by the body, inhalation is probably the most common. The very large volume of air inhaled by the average person per day means that even trace amounts of toxic materials become important and the very large surface area of the human lungs increases the chances of the materials being absorbed. Fortunately, not all the material which is inhaled is absorbed and a good proportion is exhaled. In the case of dust particles, only those particles less than about 5μm reach the lungs and there is a good chance that some will be exhaled; others will be dissolved and disposed of by the body. Dust particles, such as certain types of asbestos and similar fibrous materials, can be absorbed and have serious effects. The large surface area of the human lung allows the absorption of toxic materials to be extremely rapid. Some may also be absorbed in the mucous lining of the air passage, be brought up in the sputum and swallowed, thereby presenting additional methods of absorption. It is because inhalation of toxic materials is such a common occurrence that so much attention is paid to ventilation, monitoring of the atmosphere, and personal protective devices.

Ingestion

Ingestion of chemicals, apart from the deliberate act, is mainly due to accidental occurrences when pipetting, to contamination of food or drink, or as a consequence of inhalation as described above. All these can be avoided by taking the necessary precautions. Chemicals which are corrosive to the skin cause damage to internal organs which can result in bleeding, perforation, and deformation. The effects of toxic chemicals on the liver, kidneys, nervous system, cardiovascular system, bloodstream, bone structure, and muscles are too numerous to detail here and reference should be made to one of the standard works listed in the bibliography for further detail.

METHODS FOR DETERMINING CHEMICALS IN THE ATMOSPHERE

The literature on this subject is vast and it is only possible in the limited space available to give an indication of the rapid methods which have been introduced for examining atmospheres. For full details of methods for specific chemicals reference should be made to one of the many excellent books on the subject which are listed in the references. Standard procedures have also been specified by various national bodies typical of which are those issued by the Health and Safety Executive in the UK and the National Institute for Occupational Safety and Health Administration in the United States. The official methods must always be used if a high degree of precision is required. There are, however, several manufacturers who market devices for measuring the concentration of chemicals in air which are simple to operate and which give a reasonably accurate indication of the amount present. The principle of the method is as follows. A tube is packed with an absorbent which changes colour according to the amount of the substance being analysed which has passed through the tube. The length of coloured absorbent together with the volume of air passed through the tube is used to indicate the concentration of the substance in the atmosphere. The volume of air passed through the tube is measured by a bellows of known size and the sensitivity of the determination can be improved by increasing the volume of air. The list of chemicals that can be determined by the method is impressively large and covers most, if not all, of those in which there is a health risk. Equipment for this method of determination is marketed by several companies including Draeger, Kitagawa, and Mine Safety Appliances.

CLASSIFICATION OF FLAMMABILITY AND EXPLOSIBILITY

Many of the chemicals used in laboratories are highly flammable; admittedly the quantities of each are usually small but, unless careful control is exercised, the total volume can reach surprisingly high proportions. Several factors should be considered when comparing flammability of chemicals and these, with definitions and comments where necessary, are as follows:

Flashpoint

The flashpoint of a liquid is defined as the temperature at which the vapour from the liquid forms an ignitable mixture with air at a point close to the surface of the liquid. Various forms of apparatus for determining flashpoint of liquids at low, medium, and high temperatures are listed in the International Standards for Petroleum and its Products, which is published annually on behalf of the Institute of Petroleum by Heyden and Sons Ltd. The flashpoint may be determined with the liquid in a closed or open cup.

In the former, a small hatch is opened periodically as the sample is heated and a flame inserted to check whether the vapour from the liquid will ignite. In the UK, material with a flashpoint of less than 32°C is considered to be highly flammable as defined in the Highly Flammable Liquids and Liquefied Petroleum Gases Regulations 1972 (SI 1972 No. 917). It should be noted that it is not necessary for the source of ignition to be a flame. For example, carbon disulphide vapour can readily be ignited by a steam pipe.

Auto-ignition Temperature

This is the minimum temperature at which the substance will ignite in air without coming into contact with an external source of energy.

Lower and Upper Explosive Limits

The lower explosive limit is defined as the minimum concentration of the substance in air of normal composition which will burn when an ignition source is present. Below this concentration the mixture is too weak to burn.

The upper explosive limit marks the concentration above which the mixture of air and compound will not burn, when it is contacted with an ignition source. The oxygen concentration is too low to sustain combustion and the products of combustion quench the flame. The lower explosive limit of a fairly wide range of compounds lies between 1 and 3 per cent. Examples of typical explosive limits are given in Table 5.2. It is stressed that these are for mixtures of the substances with air of normal composition and they can vary if the oxygen content of the air deviates from normal. Clearly the vapour pressure of a flammable substance is of importance as, in the case of volatile liquids, the lower explosive limit may be reached at room temperature. A knowledge of this information will enable the correct precautions to be put into operation when dealing with known and unknown substances.

A wide range of instruments for detecting explosive atmospheres is available commercially. The principle of one type is based on the fact that combustion of the gas on an electrically heated filament will raise the temperature and hence the resistance of the filament. The change in resistance is proportional to the concentration of the combustible gas. In another version two alumina beads are used instead of the filament; one of the beads serving as a reference and the other, coated with a special catalyst, as the indicator. Exposure to flammable gas increases the resistance of the active element which provides a signal to indicate the concentration of the gas. A typical instrument for detecting flammable atmospheres is shown in Fig. 5.2.

Explosibility

This section is primarily concerned with methods for assessing the hazards of unstable substances and deals more with the evaluation of hazards associated with new substances rather than those which are well known and documented. A very considerable amount of work has been carried out on

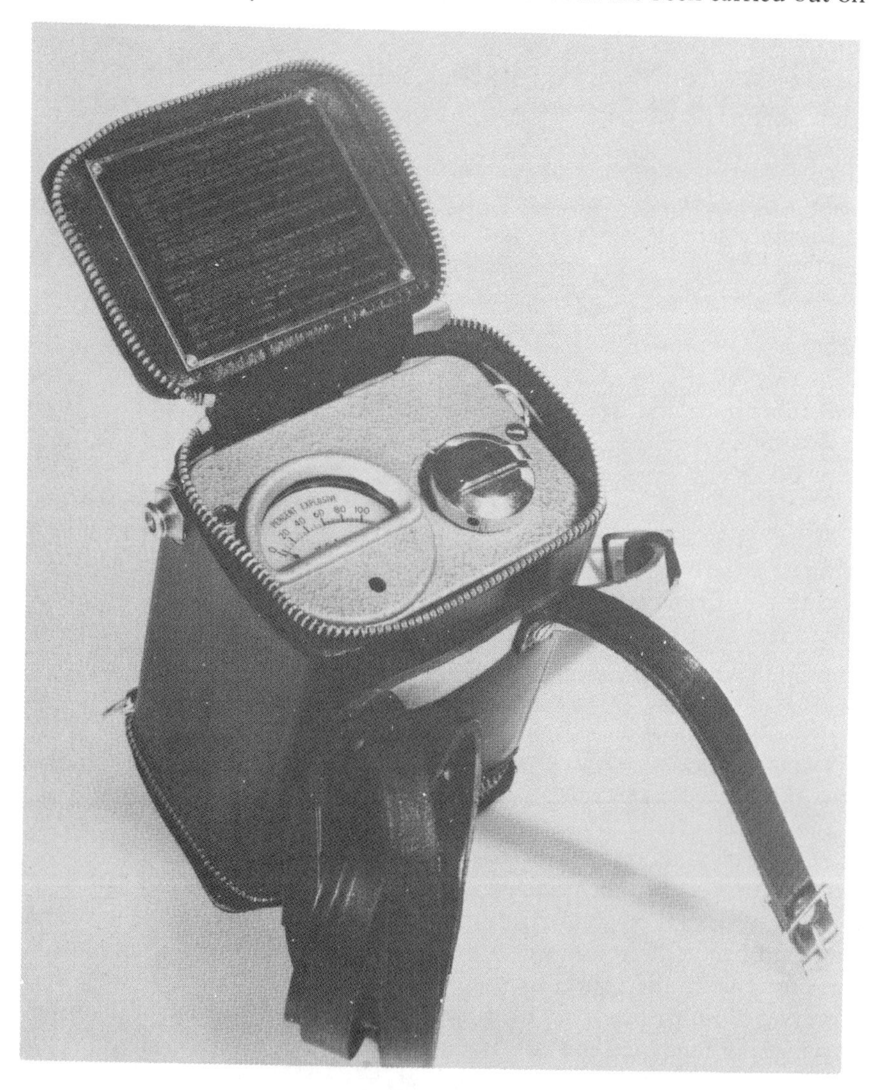

Fig. 5.2. A typical instrument for monitoring flammable or explosive atmosphere. (Photograph courtesy of Mine Safety Appliances Ltd).

Table 5.2. Examples of flashpoints, auto-ignition temperatures and limits of flammability for selected chemicals.

	Flashpoint °C		*Auto-ignition Temperature*		*Flammability Limits*	
	Closed	Open	°F	°C	Lower % v/v	Upper
Acetone	−17	−9	1000	537	2.6	12.8
iso Amyl alcohol	43	46	650	349	1.2	9.0
Benzene	−11	—	1000	537	1.4	7.1
m. Butanol	29	44	653	345	1.4	11.2
Carbon disulphide	−30	—	230	110	1.25	44.0
Cyclohexane	−17	—	500	200	1.3	8.0
Di ethyl ether	< −29	—	356	180	1.85	48.0
Ethanol 95%	14	—	—	—	—	—
Ethanol 99/100%	12	—	738	392	3.3	19.0
Glycerol	160	17>	739	392	—	—
Methanol	10	16	867	464	7.3	36.0
n-octane	13	—	428	220	1.0	3.2
Pyridine	20	—	900	482	1.8	12.4
Toluene	4	>	966	519	1.3	7.0
O-xylene	17	24	867	464	1.1	6.4
p-xylene	25	39	984	529	1.1	7.0

this subject, and it is only possible to mention here a few brief details of the methods that are in use. The first point to emphasize in work of this type is that small samples must be used to avoid the possible serious consequences of a large explosion.

The first important property to investigate is the sensitivity of the material, and both thermal and mechanical methods have been used to initiate decomposition. With the former the substance is heated and the temperature at which decomposition occurs recorded. The earlier methods consisted of dropping the material on to a hot plate and recording what happened. Recent methods include placing the material in a hypodermic needle and heating it by a condenser discharge. The explosive decomposition and temperature are obtained by measuring changes in the resistance of the needle. Differential thermal analysis methods have also been used. Mechanical sensitivity tests usually consist of variations of methods for striking the sample with a falling weight. In recent versions the substance is either struck by a rifle bullet or the substance is placed inside the bullet.

Many methods have been devised for comparing detonation characteristics of substances. The principle of one method is to heat the substance in a vented bomb and determine the critical vent size for the substance. The critical vent size is that between the size that causes the bomb to explode and one that does not. Alternatively a solid propellant igniter can be used in a vented bomb and pressure transducers used to measure the rate of pressure development with time. These methods enable materials to be classified according to their ability to undergo transition to detonation.

The sensitivity of materials to direct shock initiation can be determined by one of the many "gap" sensitivity tests. In these an explosive donor is separated from the test substance by a stack of cellulose acetate cards. Experiments are carried out to find the number of cards (or thickness or gap) at which initiation occurs 50 per cent of the time. Many variations of this test have been devised and for fuller details of this and other methods for evaluating the hazards of unstable substances Chapter 15 of reference 1 should be consulted.

GENERALLY OFFENSIVE SUBSTANCES

Included in this category are chemicals with offensive smells, lachrymatory properties, and those which produce smoke.

Odour

Although it may be argued that evil-smelling substances should not really be considered in the context of safety, nevertheless they do produce unpleasant working conditions, and frequently the chemicals are also toxic. An important case in point is hydrogen sulphide, continued inhalation of which paralyses the olfactory nerve so that a person may be unaware that potentially dangerous concentrations are being inhaled. Certain evil-smelling substances may be put to good use, e.g. those which are used to taint a flammable gas which has no odour. A list containing a selection of compounds with offensive odours is given in Table 5.3.

Considerable attention is now being devoted to odour control. There are in general two ways of dealing with odours: (a) by removal or by reducing the concentration of the odorant; or (b) by changing it to produce a material with a less offensive odour. The first approach includes ventilation, dispersal, adsorption, or absorption; while the second includes oxidation by the air, or chemical oxidizing agents such as ozone, chlorine, chlorine dioxide, permanganates, etc., and irradiation. Masking agents which do not belong in either of the above categories are becoming increasingly popular. These are agents which, when mixed with odours, modify the odour to make it more acceptable. They do not react chemically with the substance

Table 5.3. Substances with offensive odours.

Compound	*Description of odour*
Hydrogen sulphide	
Mercaptans	Offensive odours in refineries
Disulphides	
4 Mercapto 4 Methyl/onton 2 ore	"Catty" odour
Crotyl mercaptan	"Skunk" odour
Carbon disulphide	"Rotten" odour
Skatole	"Faecal" odour
Phenyl isocyanide	"Penetrating and intolerable"
Cacodye oxide	"Vile"
Thiophenol	"Putrid"
Pyridine	"Burnt"
Ethyl selenomercaptan	"Foetid"
Ethyl selenide	"Putrid"
Valeric acid	"Body odour"
Butyric acid	"Rancid"
Ethyl sulphide	"Foul"
Trimethylamine	"Fishy"
Chlorine	"Pungent"
Acrylonitrile	"Onion—pungent"
Alkyl chloride	"Foslic—pungent"
Dimethyl acetamide	"Burnt—amine"
Diphenyl sulphide	"Burnt—rubbery"

causing the odour. They may be added during the course of a chemical process, be sprayed into the atmosphere, or allowed to evaporate using a wick arrangement. The latter, together with a device for adding small amounts of a chemical deodorant and disinfectant to water closets are used extensively in the home.

Lachrymators and Smokes

These compounds cause intense eye pain and a copious flow of tears, although the effects are usually temporary. The alkyl compounds, ethyl bromoacetone, bromoacetone, bromobenzyl cyanide, benzyl bromide, and α chloroacetophenone (tear gas) all have lachrymatory properties. They have been used in chemical warfare and crowd control. The most recent chemical used in crowd control is CS gas which has a more rapid action than tear gas. Its effects on the eyes and air passages are more severe but the symptoms and pain disappear after 5 or 10 minutes in the open air.

Titanium and silicon tetrachloride are typical chemicals which hydrolyse readily in moist air and produce dense white clouds. As the clouds also contain hydrogen chloride care should be taken not to inhale them.

Nerve Agents

Mention should also be made of the nerve agents developed for use in warfare but never actually used. They are generally colourless and odourless which makes detection difficult and they are readily absorbed through the skin, lungs, and eyes without any irritant effect. Even a brief exposure may be fatal. Fortunately,they are unlikely to be encountered in normal laboratory practice.

GENERAL NOTES ON WORKING WITH HAZARDOUS CHEMICALS

The following section is by no means comprehensive and for further information one of the references should be consulted. The literature on hazards associated with chemicals is now extensive but if by chance no information can be found then recourse should be made to the manufacturer. If a compound has been synthesized for the first time then some indication of its properties may be gained from its structure. This will not always be reliable and, if it is to be developed for further use, a full-scale series of tests will have to be initiated.

Chemicals which Immediately Injure the Skin

GLOVES AND GOGGLES MUST BE WORN. When working with acids and alkalis at the laboratory bench gloves must normally be worn. When extremely delicate glass apparatus is being handled it may be permissible to relax this instruction (except in the case of hydrofluoric acid). In these circumstances conditions must be strictly controlled with plentiful supplies of water available in the immediate vicinity. Goggles or other suitable eye protection must always be worn.

Strong acids

These include the common strong acids—sulphuric, nitric, hydrochloric, hydrofluoric, chromic, glacial acetic, etc. The concentrated acids are highly corrosive to the skin, destroy paper, wood, and cloth, and attack most metals. Winchesters containing them should be kept on acid resistant trays at floor level, suitably protected from accidental knocks.

Considerable heat is produced when mixing concentrated sulphuric acid with water. For dilution use an open, thin-walled vessel (beaker or basin) and slowly pour the acid into the water—not water into acid—stirring continuously. External cooling increases the rate at which acid can be added. Oleum is even more reactive and corrosive than concentrated sulphuric acid and should preferably be diluted with concentrated sulphuric acid before adding to water. Hot acid solutions in beakers should be handled with beaker tongs.

Concentrated and fuming nitric acids cause slow-healing burns on the skin which may leave scars. The vapours are poisonous, and a 2 per cent solution of sodium hypochlorite is recommended for swabbing fuming nitric acid splashes.

Hydrofluoric acid can cause serious burns which may be painless at first. Splashes should be washed off immediately with copious quantities of water, swabbed with a solution of sodium bicarbonate and treated with calcium gluconate gel. In view of the delayed action of the acid, all suspected burn cases must be referred to hospital for treatment at the earliest possible opportunity.

Strong alkalis

These include caustic soda, caustic potash, lime, and sodium peroxide. Considerable heat is produced in making aqueous solutions of some alkalis. The alkali must be added little by little to the water with stirring, until all is dissolved.

Hot water must not be added to a caustic alkali to dissolve it, as the heat developed may throw the liquid out of the reaction vessel. When handling caustic alkalis, e.g. breaking sticks of caustic soda, gloves and goggles must be worn.

Great care must be taken when opening winchesters of concentrated ammonia solution. They must be stored in a cool place to minimize pressure build-up which could result in violent ejection of liquid when the cap is removed.

Splashes of any of the above materials must be washed off immediately with soap and water or the specified antidote. Splashes in the eye must be irrigated thoroughly by using copious amounts of water. All wounds or burns must be treated by first-aiders or, if necessary, by qualified medical personnel.

Miscellaneous chemicals

Chemicals which react violently with water include titanium and aluminium chlorides, thionyl chloride, chlorosulphonic acids, sulphonyl chloride, and phosphorus chlorides. In these cases sulphur dioxide and/or hydrogen chloride may be produced, which are injurious to the lungs.

Titanium tetrachloride must be handled with particular care. Dilution with aqueous medium must be strictly controlled using ice cooling if possible and performed in a fume cupboard. If splashes occur any contaminated clothing must first be removed. The skin must be washed with copious quantities of water as quickly as possible otherwise serious scalding may follow.

Bromine can cause severe burns and is very irritating to the eyes, nose, and lungs. Skin burns must be washed with copious quantities of water and bathed with a dilute solution of ammonia or sodium thiosulphate.

Yellow phosphorus ignites spontaneously in air and must only be handled under cold water. It must not be allowed to come into contact with oil or grease. Affected parts of the skin must be treated with a 5 per cent solution of sodium bicarbonate followed by 5 per cent copper sulphate solution.

Sodium will take fire, or even explode, if allowed to come into contact with water. Residues must not be disposed of in sinks or waste bins, but should be allowed to dissolve, a small piece at a time, in methylated spirits. If sodium comes in contact with the skin, remove any adhering metal and wash with copious amounts of water. Do not store yellow phosphorus and sodium near each other, as confusion may have serious consequences.

Aluminium alkyls are highly reactive compounds which react violently with water, alcohols, acids, etc. They cause severe skin burns which are very painful and heal badly. Protective clothing consisting of safety helmet and visor, gauntlets, and apron, must be worn at all times when they are being used. Skin burns must be cleansed immediately with a high BP saturated hydrocarbon (medicinal liquid paraffin) and then treated at the ambulance room. The white smoke produced by their interaction with moist air is harmful to the lungs and a fume cupboard with an efficient draught must be used. Small quantities of aluminium alkyls may be disposed of by dilution with toluene followed by decomposition with isopropanol. Alternatively, acetone dilution followed by a water quench is equally effective.

Alkali metal hydrides react as vigorously with water as the alkali metals themselves and produce large volumes of hydrogen and heat—a dangerous combination. One rather more complex hydride which is now in fairly common use is lithium aluminium ($LiAlH_4$). This is an extremely powerful and selective reducing agent which reacts vigorously with water to produce hydrogen and heat. Even the act of pouring the solid can cause electrostatic sparks to ignite hydrogen. It is also a powerful caustic agent. An explosion has been reported when using this chemical to dry diethylene glycol dimethyl ether. This was believed to be due to caking of the hydride on the wall of the flask while distilling the ether and the development of a hot spot in the cake. Gloves and face shields must be worn when working with $LiAlH_4$ and all means of ignition removed from the vicinity of the experiment. The apparatus must also be protected by safety screens. High test hydrogen peroxide is associated with spontaneous decomposition followed by combustion, which occurs when concentrations above 65 per cent w/w come into contact with organic matter such as wood, dirt or rags. PVC gloves and goggles must be worn when using solutions of strength greater than 30 per cent w/w and absolute cleanliness of equipment must be observed. A supply of clean water under pressure must be available to dilute the peroxide if trouble occurs.

Chemicals Absorbed by the Skin

Poisoning may result from allowing certain chemicals to remain on the skin. Such chemicals include methanol, nitro and amino derivatives of benzene and toluene, e.g. aniline, mononitrobenzene, dinitrochlorobenzene, phenyldiamines, and phenols. Phenols rapidly attack the skin and cause serious burns although poisoning by absorption through the skin may occur even though the skin surface has remained intact. Inorganic compounds include lead oxides and salts, arsenic, copper, selenium and mercury compounds. The hands must be washed immediately with soap and water after using such chemicals.

Cumulative Poisons

Certain chemicals that are relatively harmless when used intermittently are harmful if constantly or regularly used, owing to cumulative absorption of small quantities. The more common of these are the elements or compounds of lead, arsenic, and mercury. Carbon tetrachloride, benzene, tetrachlorethane, and the nitro- and amino- derivatives of benzene and toluene are dangerous if breathed continuously.

Inhalation of carbon tetrachloride vapour under certain circumstances can be fatal. It is strongly recommended that its use be forbidden and that perchlorethylene be used in its place. Perchlorethylene is much less toxic and has very similar properties to carbon tetrachloride.

Mercury metal is particularly toxic and has a measurable vapour pressure at laboratory temperatures. Even at 15°C, the concentration of saturated mercury vapour in air is at least 70 times the permissible concentration. Any spilled mercury must be picked up at once using a capillary tube, suction, and a collection flask in the suction line. If it is necessary to leave a surface of mercury exposed to the atmosphere it must be covered with a layer of water or other non-toxic liquid.

FLAMMABLE LIQUIDS AND VAPOURS

No more than a day's supply of flammable liquids may be kept in each laboratory. Many liquids in general use readily catch fire and some are exceedingly flammable. Flammable liquids must not be poured near naked flames nor must they be heated over a flame, except for specification test purposes. Large flasks containing flammable liquids should be placed in a pan of capacity greater than the flask so that in the event of breakage the contents are safely trapped. Use must be made of steam and electric heating. Take particular care that apparently unlit Bunsen burners are not burning back at the jet. Electrical apparatus such as hotplates, water baths, iso-mantles, etc. for use where naked flames are prohibited, must be guaranteed as flameproof. Do not place bottles of volatile flammable liquids in direct sunlight. Do not smoke when working with flammable liquids. Most of the lacquer solvents are low-flash liquids which easily take fire in the presence of flames or sparks.

Frequently, flammable organic liquids are used in heating baths. These must only be used as a last resort if isomantles or non-flammable liquids are unsatisfactory. Care must be taken to ensure that the operating temperature does not exceed 50°C below the closed cup flashpoint of the liquid. Free space in oil baths should be sufficient to hold the contents of the vessel, in the event of a fracture, in order to localize any resulting fire.

Carbon disulphide must be used with particular care since a steam pipe or electric light bulb can ignite it. It burns to form sulphur dioxide and is itself poisonous.

Some highly flammable liquids in common use are ethyl ether, benzene, and low boiling petroleum ethers. Ethyl ether requires particular care. On account of its low boiling point and the high flammability of its vapour which tends to creep along the bench or water gully, no flames should be permitted on the same bench.

Other liquids whose vapours are readily flammable are: methanol (methyl alcohol); ethanol (ethyl alcohol, methylated spirit); petroleum (ligroin, kerosene or paraffin); acetone, toluene, xylene, solvent naphtha, white spirit, low boiling esters.

Dusts and Fumes

Poisoning may occur from the inhalation of dusts and fumes produced in operations such as crushing, grinding, screening, and riffling. Metal fumes, lead, chromium, cadmium, phosphorus, selenium, mercury, beryllium, vanadium compounds, etc. are particularly toxic. Silica, siliceous dusts, and asbestos can also cause damage to the lungs.

Work involving the above materials must be conducted in an efficient fume cupboard or, where this is not practicable, suitable protective masks must be worn.

Peroxides and Peroxides of the Ethers

Danger of fire is not the only risk when using ethers. Diethyl, isopropyl, and higher ethers, when exposed to air and bright light, may form unstable peroxides that are explosive upon evaporation to dryness. A small amount should be tested for peroxides before distilling any quantity.

Supplies should be stored away from light, in amber-coloured bottles and for the shortest possible period. Beware of ethers which have been standing for some months, particularly if in ordinary clear glass bottles with a considerable volume of air over the liquid.

The presence of peroxides in ether may be detected by shaking a small volume with an equal volume of 2 per cent potassium iodide solution and a few drops of dilute hydrochloric acid. Liberation of iodine (brown coloration) confirmed by the blue coloration produced with starch solution, indicates the presence of peroxides. They may be removed by shaking the ether with a strong solution of ferrous sulphate (60 g ferrous sulphate, 6 ml concentrated sulphuric acid dissolved in 110 ml water) or with an aqueous solution of sodium sulphite.

The most common organic peroxides are benzoyl peroxide, methyl ethyl ketone peroxide, acetyl peroxide, and peracetic acid. All these compounds are dangerous if handled carelessly and are frequently mixed with an inert compound such as chalk to make them less dangerous to handle (phlegmatized). They are liable to react violently with combustible materials and friction can bring about fires or explosions. They should not be stored with flammable or easily oxidized materials.

Polyvinyl Chloride (PVC)

Although PVC does not burn, at high temperatures it decomposes to form hydrogen chloride which is highly corrosive and a respiratory irritant.

Perchloric Acid: Explosion Hazard

Perchloric acid, in the presence of easily oxidizable organic or inorganic matter, can produce dangerous explosions and its use must be strictly and directly supervised by a senior member of the laboratory staff.

The constitution and properties of any material must be taken into consideration before treatment with perchloric acid. Generally, mixtures of

72 per cent perchloric acid and nitric acid may be used safely for the destruction of organic matter. The nitric acid must be added first and the operation carried out with all safety precautions.

Solutions containing alcohol, glycerol, or other substances that form esters, must not be heated with perchloric acid or perchloric mixtures, since esters of perchloric acid are highly explosive. Some inorganic materials such as hypophosphites and antimony compounds tend to form explosive mixtures with hot perchloric acid and a large excess of nitric acid should be present during their oxidation.

Evaporation with perchloric acid must be carried out in a fume cupboard which is kept clean and free from combustible materials; alternatively a fume eradicator may be used.[2] The draught must effectively prevent condensation of the acid vapours and the fume cupboard window must be down during the evaporation period.

Perchloric acid must not be allowed to come into contact with wooden shelves or benches, and containers must stand on glass or porcelain dishes. Any acid spillage must be neutralized immediately with sodium carbonate and washed with copious quantities of water before being mopped up. In all work involving perchloric acid, rubber gloves, eyeshields, or goggles and safety screens must be used.

Magnesium perchlorate ("Anhydrone") is an extremely efficient drying agent but it is not recommended for the drying of organic liquids because of risk of the formation and explosion of organic perchlorates. It is claimed that the explosions which have occurred are due to the presence of traces of perchloric acid in the magnesium perchlorate and that the pure agent is safe. The safest course is to use an alternative drying agent or combination of agents.

Nitrocellulose

Industrial nitrocellulose is usually damped with at least 30 per cent of damping agent (water, industrial methylated spirit, butanol or isopropanol). In such a condition it presents a hazard no greater than that of other flammable solvents, etc. used in the lacquer industry. It is very important, however, to prevent the nitrocellulose from becoming dry, as in this condition it is very dusty and liable to settle and accumulate in niches. Dry industrial nitrocellulose is very easily ignited by sparks, friction, impact, flame, or static electricity, and burns with great speed and intense heat.

The following points relating to the storage and handling of industrial nitrocellulose should be noted:

1. Nitrocellulose should be stored in a suitable place away from rooms where manufacturing processes are in operation. The building should be constructed of fire-resisting material and preferably used for no other purpose than for storing nitrocellulose.
2. No open light or fire should be permitted near this building. The building should be kept locked and entrance forbidden to unauthorized persons. Means of ignition and anything likely to cause fire or explosion should be prohibited within the storage building. These stipulations also apply to the buildings in which nitrocellulose is processed.
3. The material should not be stored in a building into which direct rays of the sun enter.
4. Radiators should be avoided if possible. If their use is imperative they should be heated only by means of low-pressure steam and be in such a position that the nitrocellulose containers cannot be placed against them.
5. Adequate means of escape from the building should be provided, as well as suitable means of extinguishing fire.
6. Containers should not be stored in tiers.
7. Containers should be handled carefully to avoid sparks on contact with steel or concrete.
8. No undue force should be used in opening containers, and tools used for the purpose should be of brass or other non-sparking material. (Suitable tools may be obtained from Imperial Chemical Industries plc.)
9. The weighing of nitrocellulose should be isolated so that ignition dangers consequent on the formation of dry nitrocellulose dust may be kept away from other operative work.
10. The walls and roofs of storage and processing buildings should be washed down periodically to remove dry nitrocellulose dust.
11. The lids of nitrocellulose containers should be replaced as soon as possible in order to prevent the evaporation of the damping medium.
12. The amount of nitrocellulose in any manufacturing building should be kept as small as practicable. Any waste should be collected at once, deposited in a suitable receptacle, wetted with water and collected at the end of the day for burning on a wood fire in an isolated and safe area. Such waste should never be disposed of in a boiler fire.
13. Containers should be inverted at regular intervals (say weekly) during storage in order to maintain even distribution of the damping medium.
14. Surplus nitrocellulose should be disposed of as soon as possible and not stored indefinitely.

Methyl Methacrylate Monomer

This material is flammable and can polymerize rapidly with generation of heat. This will result in rupture of the container if sealed with a screw type stopper. Polymerization is normally prevented by the addition of hydroquinone, but if it is essential to work with unstabilized material it should be stored in a cool place in a container with a loosely fitting plug instead of a screw type stopper.

Alternative stabilizing agents such as Topanol A or the methyl ether of hydroquinone cannot be tolerated.

Carcinogenic Substances

In almost every country there are regulations regarding the handling and use of carcinogenic substances and clearly these must be complied with and also referred to where necessary for additional information. The UK's Carcinogenic Substances Regulations of 1967 are typical and contain the requirements mentioned below which are indicative of those elsewhere.

Use of the following substances and their salts is prohibited except where exemption is granted in writing by the Chief Factory Inspector: beta naphthylamine, benzidine, 4-aminodiphenyl and 4-nitrodiphenyl. The risk factor with these compounds is high and alternative compounds must be used, where possible.

The following substances and their compounds are designated Controlled Substances: alpha-naphthylamine, ortho-tolidine, dianisidine, dichlorbenzidine, auramine, and magenta. These substances may be used subject to certain controlling conditions, e.g. compulsory six-monthly medical examination and registration of periods of working with such substances. It is strongly recommended that alternative materials to these be used where possible.

Special regulations have been issued under the Factories Act for asbestos. These are known as the Asbestos Regulations 1969 and should be referred to, if necessary, for further information. The principal use for asbestos in laboratories is for thermal insulation and alternative means must be used where possible. A suitable substitute for asbestos in lagging is asbestos-free magnesia known as 85 per cent supermagnesia. A PVC-coated scrim cloth known as Darcode is an acceptable substitute for asbestos cloth in some applications. The greatest hazard exists with blue asbestos (CROCIDOLITE) and this must never be used. The regulations may be summarized as follows:

1. Where possible, work with asbestos should be carried out using exhaust ventilation so that asbestos dust is excluded from the air in the laboratory.
2. Where item 1 is impracticable approved respiratory equipment and protective clothing must be worn. This will be obtained from the Safety Store and returned for cleaning purposes immediately after use.
3. The working area must be kept clean. This must be done in such a way that asbestos dust cannot enter the atmosphere. Vacuum cleaning is recommended. If cleaning cannot be done by a dustless method then protective clothing and protective respiratory equipment must be worn. When asbestos cloth is being cut, the material should be wetted in the area to be cut.
4. All loose asbestos must be stored and transferred in closed receptacles which prevent the escape of dust into the atmosphere when not in use.

Fluoropolymers

This category includes polytetrafluoroethylene, polyvinylfluoride, polyvinylidene fluoride, polychlorotrifluoroethylene, and copolymers of hexafluoropropylene.

Any laboratory in which these materials are likely to be present in the atmosphere should observe a "No Smoking" rule since they can decompose in a cigarette flame to liberate hydrofluoric acid (and in the case of polytetrafluoroethylene and its copolymers, perfluoroisobutylene which is also very poisonous).

HANDLING OF GASEOUS POISONS AND IRRITANTS

Irritants include the vapours from many acids (hydrochloric, hydrofluoric, nitric), sulphur chlorides and bromine and especially diphenylchloro and cyanoarsines (used in war gases) as well as substances which occur in the gaseous state such as chlorine, sulphur dioxide, phosgene, and nitrogen peroxides.

Phosgene, nickel carbonyl, and nitrous fumes have a delayed action upon the lungs producing initially slight coughing and sickness. This passes off and is followed after a few hours by complete collapse. Note that since phosgene is generated when chlorinated hydrocarbons react with a hot surface, smoking is not allowed when using them.

Carbon monoxide is odourless and gives no sense of irritation, consequently there is no warning of its presence. Experiments should preferably be carried out in the open air but if this is not possible, they should be carried out in well-ventilated fume cupboards or hoods. It is also advisable to instal a carbon monoxide detector which indicates its presence and operates an alarm when the concentration reaches a dangerous level. One source of supply is the Union Industrial Equipment Corporation, 40 Beach Street, Port Chester, NY, USA.

Hydrogen sulphide is more dangerous than is often supposed, as paralysis of the sense of smell (anosmia) may occur.

Hydrogen selenide is a highly poisonous gas which resembles arsenic in its physiological activity and which is almost as toxic as hydrogen cyanide.

The vapours and gases referred to above, and any others which are classified as toxic, must not be allowed to escape into the air of the laboratory. Work involving them must be carried out in a fume cupboard with an efficient draught. In cases where the fume is particularly toxic ensure that no persons are near the fume chamber discharge point. Where highly toxic gases which cannot be detected by smell are used in quantity, or over a long period, a respirator fitted with a canister suitable for the particular respirator risk should be worn. When concentrations greater than 1 per cent by volume are anticipated or known to be present then self-contained breathing apparatus should be used. In certain cases a chemical method for detecting the gas can be used provided that immediate indication is given when the concentration reaches the dangerous level. In such cases where the gas is readily detected the respirator need not be worn continuously but must be readily available. This exclusion does not apply where highly toxic substances are or may be present in the atmosphere. A suitable danger warning notice should be displayed near apparatus where toxic or irritant gases are being used or given off.

If any of these gases are suddenly released in the laboratory through breaking or bursting of a container there must be immediate evacuation of all unprotected personnel in the area and no return to the laboratory until it has been thoroughly ventilated and certified as free from gas, except for those wearing appropriate protective equipment.

DISPOSAL OF CHEMICALS

EXPERT ADVICE SHOULD BE SOUGHT IN CASES WHERE THE EXACT PROCEDURE FOR DISPOSING OF HAZARDOUS CHEMICALS IS NOT KNOWN OR WHERE THE ACTUAL NATURE OF THE MATERIAL IS UNKNOWN.

In the space available it is impossible to discuss each particular case in detail. It is, however, occasionally necessary to dispose of chemicals with dangerous properties and in some cases the nature of the chemical may be unknown because of the label becoming detached from the bottle. In these circumstances ignorance of correct procedure can lead to personal injury.

In addition to the normal fire and explosion hazards of flammable liquids and the expected hazards of toxic materials, certain other hazards must be anticipated in waste disposal. Some materials are corrosive to drainage piping, some react violently with water or other chemicals; and others, while relatively non-hazardous in themselves, adversely affect sewage disposal systems.

Chemicals poured down the drain should be non-toxic or in concentrations below the threshold limit. The concentrations which may be disposed of in this way are usually subject to control and advice of the controlling authority should be obtained in cases where it is suspected that pollution may occur. It may be necessary to pretreat the chemical or dispose of it by other means, e.g. by burning or by slowing down the rate of decomposition. There are several useful publications dealing with the handling and disposal of dangerous chemicals.[3,4,5]

DISPOSAL OF AEROSOL CANS

Because aerosol cans are sealed with a valve, even when exhausted they can be a potential danger if heated, incinerated, or pierced. They should therefore be placed in a separate marked container for collection by the refuse disposal service.

REFERENCES

1. "Guide for Safety in the Chemical Laboratory" (1972), 2nd edn. Manufacturing Chemists Assoc., Van Nostrand Reinhold Co.
2. Vogel, A. I. (1975). "Quantitative Inorganic Analysis", 3rd ed., p. 230, Longman, London.
3. Gaston, P. J. (1970). "The Care, Handling and Disposal of Dangerous Chemicals", The Institute of Science Technology, Northern Publishers (Aberdeen) Ltd., 11 Albyn Terrace, Aberdeen, UK.
4. "Laboratory Waste Disposal Manual" (1969), published by Manufacturing Chemists Association, Washington DC, USA.
5. "How to deal with Spillages of Hazardous Chemicals", B.D.H. Wall Chart, obtainable from B.D.H. Chemicals Ltd., Poole, Dorset, UK.

FURTHER READING

American Conference of Government Hygienists (1979). Documentation of Threshold Limit Values for Substances in the Work Room Air.

"Applying TLV's" (1980). *Occupational Health*, **32,** No. 6, 301.

Berman, E. (1980). "Toxic Metals and their Analysis", Heyden, London.

Bretherick, L. (1975). "Handbook of Reactive Chemical Hazards", Butterworths, London.

Bretherick, L. ed. (1981). "Hazards in the Chemical Laboratory", 3rd edn., The Royal Society of Chemistry, London.

Browning, E. (1965). "Toxicity and Metabolism of Industrial Solvents", Elsevier, Amsterdam.

Browning, E. (1953). "Toxic Solvents", Edward Arnold, London.

Browning, E. (1961). "Toxicity of Industrial Metals", Butterworths, London.

"Chemical Safety Data Sheets", Manufacturing Chemists Association Inc., Washington.

"Codes of Practice for Chemicals with Major Hazards", Chemical Industries Association, UK.

Cooke, J. (1979). "Perchloric Acid, the Dangers of Contamination in the Laboratory", *Health & Safety at Work*, **54.**

"Dangerous Substances, Guidance on Dealing with Fires and Spillages" (1972). HMSO, London.

Deichmann, W. B. and Gerard, H. W. (1973). "Toxicity of Drugs and Chemicals", Academic Press, London and New York.

Dreisbach, R. H. (1971). "Handbook of Poisoning: Diagnosis and Treatment", Lange Medical Publications, Los Altos, C.A., USA.

Eagers, R. Y. (1969). "Toxic Properties of Inorganic Fluorine Compounds", Elsevier, London.

"Effects of Exposure to Toxic Gases, First Aid and Medical Treatment" (1970). Matheson Gas Products, USA.

Elkins, H. B., 2nd edn. (1959). "The Chemistry of Industrial Toxicology", John Wiley & Sons, London and New York.

"An Encyclopaedia of Chemicals and Drugs", 9th edn., The Merck Index, Merck & Co.

Fairchild, E. J. (1978). "Suspected Carcinogens", Castle House Publications Ltd.

"Fisher Manual of Laboratory Safety" (1972). Fisher Scientific Company, Pittsburgh, USA.

"Flash Points" (1962). The British Drug Houses Ltd., Poole, Dorset, UK.

Gaston, P. J. (1970). "The Care, Handling and Disposal of Dangerous Chemicals", Institute of Science Technology, Northern Publishers (Aberdeen) Ltd., 11 Albyn Terrace, Aberdeen, UK.

Gemert, L. J. Van and Nettenbreijer, A. H. (1977). "Compilation of Odour Thresholds in Air and Water", National Institute for Water Supply, Voorburg; Central Institute for Nutrition and Food Research, TNO Zeist, The Netherlands.

Gosselin, R., *et al.* (1976). 4th edn., "Chemical Toxicology of Commercial Products", Williams & Wilkins Co.

Goyer, R. A. and Melhman, M. A. (1977). "Toxicology of Trace Elements", John Wiley & Sons, London.

Green, M. E. and Turk, A. (1978). "Safety in Working with Chemicals", Macmillan Publishing Co. Inc., New York and Collier Macmillan Publishers, London.

Harvey, B. (1980). "Handbook of Occupational Hygiene", Kluwer Publishing Co., London.

Handley, W., ed. (1977). "Industrial Safety Handbook", McGraw-Hill Book Company (UK) Ltd.

"Handling Chemicals Safely" (1980). 2nd edn., Dutch Association of Safety Experts, Dutch Chemical Industry Association, Dutch Safety Institute.

Harvey, B. and Murray, R. (1958). "Industrial Health Technology", Butterworths, London.

Hawley, G. A. (1977). "Condensed Chemical Dictionary", Van Nostrand Reinhold Co., New York.

"Highly Flammable Liquids and Liquefied Petroleum Gases, Guide to the Regulations" (1972). HMSO, London.

Hilado, J. C. and Clark, S. W. (1972). "Auto-ignition Temperatures of Organic Chemicals", *Chemical Engineering* **75–80,** Sept. 4th.

Hunter, D., 6th edn. (1978). "The Diseases of Occupations", Hodder & Stoughton.

"Industrial Dermatitis", 2nd edn. (1960). Chemical Industries Association, London.

"Industrial Nitrocellulose" (1961). Mond Division, Imperial Chemical Industries

plc, London.
"Industrial Solvents, Flammable Liquids and Low Melting Point Solids" (1965). Fire Protection Association, Aldermary House, Queen Street, London, UK.
"International Labour Office Encyclopaedia of Occupational Health & Safety" (1971). 2 Vols, International Labour Office, Geneva.
"Laboratory First Aid Wall Chart" and "Spillages of Hazardous Chemicals Chart", B.D.H. Chemicals Ltd., Poole, Dorset, UK.
"Laboratory Waste Disposal Manual" (1969). Manufacturing Chemists Association, Washington, USA.
Mann, C. A. (1969). "Safety in the Chemical Laboratory LVIII, Science Experiment Safety in the Elementary School", *J. Chem. Educ.* **46,** Part 5, A347–353.
Marsden, C. and Mann, S. (1963). "Solvents Guide", Macmillan, London.
Martindale, "The Extra Pharmacopeia", 27th edn. (1977). Pharmaceutical Press, London.
Matthew, H. and Lawson, A. A. H., 4th edn. (1979). "Treatment of Common Acute Poisonings", Churchill Livingstone, London.
Paget, G. E., ed. (1979). "Topics in Toxicology—Good Laboratory Practice", MTP Press Ltd, London.
Palty, F. A., ed. (1963) "Industrial Hygiene & Toxicology, Vol. II", 2nd edn., Interscience, New York and London.
"Phosgene" (1975). Codes of Practice for Chemicals with Major Hazards, The Chemical Industry Safety and Health Council of the Chemical Industries Association, London.
Plunkett, E. R. (1976). "Handbook of Industrial Toxicology", Hayden, London.
Porter, W. E., Ketchen, E. F. *et al.* (1978). "Health Considerations Relative to the Use of Solvents in the Chemistry Laboratory", *Energy Res. Abstr.* **3,** Part 20, Abstract No. 48907.
"Proceedings of the International Symposium on Maximum Allowable Concentrations of Toxic Substances in Industry" (1961). Butterworths, London.
"Protection of the Eyes" (1963). British Chemical Industry Safety Council, London.
"Raw Materials Safety Data Handbook" (1976). Paint Makers Association of Great Britain Ltd, Prudential House, Wellesley Road, Croydon, UK.
"Registry of Toxic Effects of Chemical Substances" (1978). National Institute for Occupational Safety and Health, US Department of Health, Education and Welfare, Washington, DC.
"Safety in the Chemical Laboratory" (1966). May and Baker Ltd, Dagenham, Essex, UK.
"Safety in Chemical Laboratories and in the Use of Chemicals" (1970). Imperial College of Science and Technology, London.
Sax, N. I., 5th edn. (1979). "Dangerous Properties of Industrial Materials", Van Nostrand Reinhold Company, New York.
Stalzer, R. F., Martin, J. R. and Railing, W. E. (1967). Safety in the laboratory. *In* "Treatise on Analytical Chemistry", Part III, Vol. 1 (I. M. Kolthoff, P. J. Elving and F. H. Stross, eds). Interscience Publishers, New York.
"The Storage of Highly Flammable Liquids", Chemical Safety/2 January 1977,

Health and Safety Executive, HMSO, London.
Steere, N. V. (1971). "Handbook of Laboratory Safety", C.R.C. Press Inc., Boca Raton, Florida, USA.
"Threshold Limit Values for 1980". Guidance Note EH 15/80, Health and Safety Executive, HMSO, London.
"Toxic and Hazardous Industrial Chemical Safety Manual" (1976). The Information Institute, Tokyo.
Weart, R. C., ed., 61st edn. (1980). "Handbook of Chemistry and Physics", The Chemical Rubber Publishing Co, Cleveland, Ohio, USA.
Zabetakis, M. G. (1964). "Flammability Characteristics of Combustible Gases and Vapours", US Bureau of Mines, Bulletin 627, Washington DC.

6

Electromagnetic Radiation Hazards

The literature on controlling hazards due to radiation and lasers is now immense and in this chapter it is only possible to deal with the subject in broad terms. For more detailed information the reader is referred to the references at the end of this chapter which includes a selected list of authoritative bodies who have laid down legislation on ionizing radiations.

TYPES OF RADIATION

The electromagnetic spectrum is shown in Fig. 6.1 and contains radiation of wavelengths ranging from 10^{-12}cm to 10^{4}cm. The spectrum ranges from cosmic and gamma radiation which covers the range 10^{-12} to 10^{-8}cm, through X-radiation, ultraviolet, visible, infra-red, and microwave zones to radio frequencies of 10^{2} to 10^{4} cm. Hazards associated with each of these will be discussed in the following sections starting with cosmic radiations.

COSMIC RADIATION

This is radiation received by the earth from space which, together with natural radioactivity, represents the background radioactivity to which the human race has always been exposed. The average background radiation is approximately 5×10^{-4} Sievert per year (for definition of Sievert see p. 100 but in some parts of the world levels as high as 0.12 Sievert have been recorded.

IONIZING RADIATIONS

Main Types

Ionizing radiations consist of electromagnetic or corpuscular radiation capable of producing ions directly or indirectly in its passage through

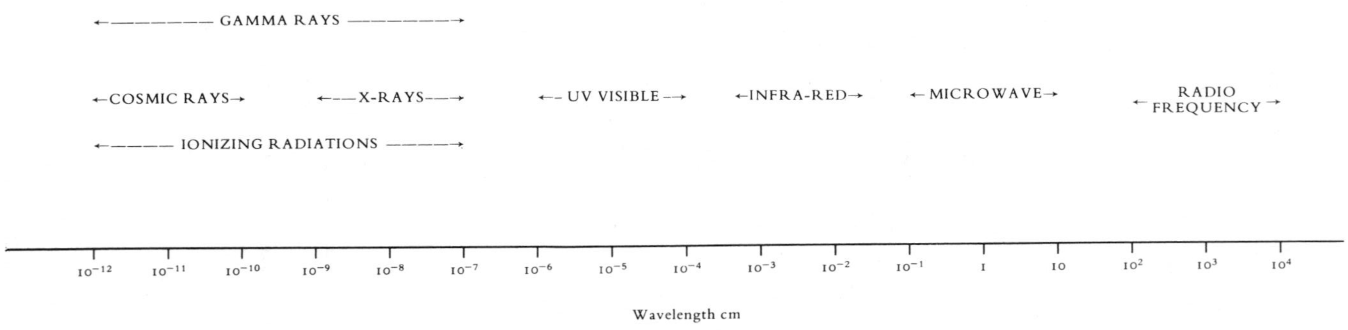

←——— GAMMA RAYS ———→

←COSMIC RAYS→ ←—X-RAYS—→ ←— UV VISIBLE —→ ←INFRA-RED→ ← MICROWAVE→ ← RADIO FREQUENCY →

←——— IONIZING RADIATIONS ———→

10^{-12} 10^{-11} 10^{-10} 10^{-9} 10^{-8} 10^{-7} 10^{-6} 10^{-5} 10^{-4} 10^{-3} 10^{-2} 10^{-1} 1 10 10^{2} 10^{3} 10^{4}

Wavelength cm

Fig. 6.1. The electromagnetic spectrum.

matter. Notice that corpuscular radiation, e.g. α-particles and neutrons, have been included. The characteristics of the main species of ionic radiations are described below:

X-rays

X-rays are produced when a fast moving stream of electrons or charged particles interacts with matter. Electrons are ejected from inner shells of the atoms of the target and, as electrons from outer shells fall back to take their place, energy is released in the form of X-rays. X-rays are normally produced in a vacuum tube (see Fig. 6.2) in which electrons from a heater filament are accelerated towards a water-cooled anode target. As the accelerating voltage can be varied, usually from 10 Kv to 100 Kv, the wavelength of the X-rays which are produced can be varied from the short wavelength, high energy or "hard" type to "soft" X-rays of lower energy and longer wavelength. X-rays are now used extensively in laboratories for structural and chemical analysis.

It should be noted that X-rays are not only produced in equipment designed for their generation but may be present in equipment in which a fast moving electron beam or stream of charged particles comes into contact with matter. X-rays therefore are generated in instruments such as

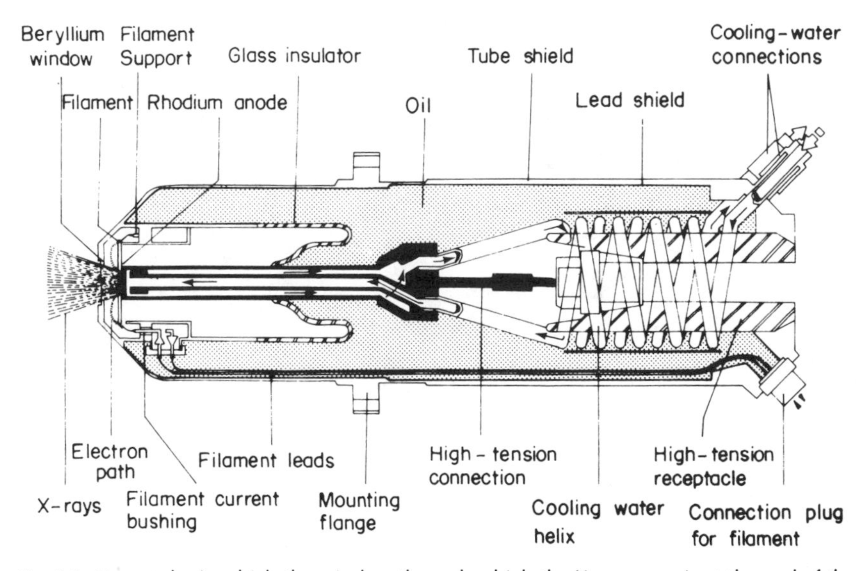

Fig. 6.2. X-ray tube in which the window through which the X-rays pass is at the end of the tube. The electrons emitted by the filament strike the rhodium anode and cause X-rays to be generated which pass through the beryllium window. (Photograph courtesy of N.V. Philips Gloeilampen Fabriken).

electron microscopes and mass spectrometers and care must be taken to ensure that harmful emission of X-rays does not take place from those instruments. The domestic television set comes within this category but the X-radiation is relatively soft and simple precautions suffice to make the set safe.

Gamma-rays

These are physically similar to X-rays but of shorter wavelength. They are emitted when most radioactive changes take place and usually accompany α- and β-radiation.

α-rays

These consist of streams of α-particles or helium atomic nuclei travelling at high speed. They are emitted by certain heavy element radioactive nuclides, have little penetrative power but great powers of ionization.

β-rays

β-rays are actually electrons emitted by radioactive nuclides. They are more penetrative than α-rays but have moderate powers of ionization.

Neutrons

Neutrons are emitted spontaneously by certain radioactive materials and may also be produced by bombarding a target containing deuterium or tritium with a fast-moving stream of deuterium or tritium atoms. They are uncharged but are capable of initiating radioactive disintegrations when they react with other atoms.

Bremsstrahlung

Bremsstrahlung is a form of X-radiation produced by the interaction of β-rays with matter.

Positrons

Positrons are similar to electrons but carry a positive charge and are very short lived.

Sealed and Unsealed Sources

A sealed source may be defined as a source of radiation sealed in a container (but not solely for the purpose of transport, storage or disposal) or bonded within a material such that it is impossible to come into contact with it. This includes X-ray tubes but does not include any nuclear fuel element or any radioactive substance inside a nuclear reactor.

An unsealed source is any other radioactive substance.

Definitions

In considering definitions of radioactivity SI units will be used and the relation of these to the earlier units will be included.

Activity

A radioactive material undergoes disintegration and the rate at which it disintegrates is called its activity. The SI unit of activity is called the Becquerel (Bq) which is defined as a rate of one disintegration taking place per second, i.e.

$$1\,\text{Bq} = 1\,\text{S}^{-1}$$

The previous unit for defining activity was the Curie (Ci) which was defined as follows:

$$1\,\text{Ci} = 3.7 \times 10^{10} \text{ disintegrations per second}$$
$$\text{or } 1\,\text{Ci} = 3.7 \times 10^{10}\,\text{Bq}$$
$$\text{or } 1\,\text{Bq} = 2.7027 \times 10^{-11}\,\text{Ci}$$

The value of the Curie was originally chosen because it is approximately equal to the number of disintegrations per second occurring in 1 g of radium in equilibrium with its daughter products.

The rate of disintegration of a single radioactive nuclide is constant and is usually expressed in terms of the time taken for the activity to decrease to one half of the initial value or, in other words, the half life of the nuclide. Half lives vary enormously from 8×10^{-19} seconds in the case of $_5B^N$ to 1.28×10^9 years for $_{19}K^{40}$.

Röntgen

Exposure to radiation is measured in terms of this unit. It is defined as the quantity of X- or α-radiation for which the associated corpuscular emission per cm³ of air produces ions that carry 1 electrostatic unit of electricity of either sign.

Absorbed dose (D)

The absorbed dose is defined as the quotient of $d\bar{\varepsilon}$ by dm where $d\bar{\varepsilon}$ is the mean energy imparted by ionizing radiation to matter in a volume element and dm is the mass of matter in that volume element. The unit of absorbed dose is the Gray (Gy) defined as:

$$1\,\text{Gy} = 1\,\text{J}\,\text{Kg}^{-1}$$

The amount of energy absorbed in exposure to 1 röntgen is dependent on factors such as the energy of radiation and the nature of the absorbing material and hence the absorbed dose was measured by a different unit, the rad (röntgen absorbed dose) defined as an energy absorption of 100 erg g^{-1} ($10^{-2}\,\text{J}\,\text{Kg}^{-1}$):

$$1\ \text{rad} = 10^{-2}\ \text{Gy}$$
$$1\ \text{Gy} = 100\ \text{rad}$$

Dose equivalent (H)

For protection purposes, the term dose equivalent has been introduced to take into account the quality factor of the radiation (e.g. neutrons and protons produce greater biological damage for a given dose than X-, α-, or β-rays), the duration of exposure and the dose rate. The dose equivalent is expressed in terms of rem (rad-equivalent-man) and maximum permissible doses are expressed in Sievert (Sv):

$$1\ \text{rem} = 10^{-2}\ \text{Sv}$$
$$1\ \text{Sv} = 100\ \text{rem}$$

For all X-, α- or β-rays the dose equivalent in rem is numerically equal to the dose in rads.

Maximum Permissible Doses

The data below are taken from the Council Directive of 15 July 1980 (80/836 Euratom) published in the Official Journal of the European Communities of 17 September 1980 (L246 volume 23).[1] This Directive falls to be implemented by member States of the Community by 3 December 1982. This document should be consulted for more detailed information as it is only possible to include the salient points in this section. The UK Health and Safety Executive issued late in 1981 a consultative document containing a draft schedule of dose limits, draft regulations, and draft Approved Codes of Practice, which should also be consulted.

Every activity resulting in an exposure to radiation should be justified by the advantages that it produces; all exposures should be kept as low as reasonably achievable and the dose limits should not be exceeded. Exposed workers are defined as persons subjected, as a result of their work, to an exposure liable to result in annual doses exceeding one-tenth of the annual dose limits for workers. The information in the Directive may be summarized as follows:

Workers under the age of 18 years may not be assigned to any work which would result in their being exposed workers.
Nursing mothers shall not be employed in work involving a high risk of radioactive contamination; if necessary, a special watch will be kept for bodily radioactive contamination.

Table 6.1.

	Dose Limit in mSv	
	Exposed Workers	*Member of the Public*
Whole body exposure	50	5
Women of reproductive capacity, the dose to abdomen shall not exceed	13	—
Pregnant women, dose to foetus should be kept as low as possible and in no case should exceed	10	—
Partial body exposure:		
Effective dose	50	5
Average dose in each organ or tissues per year	500	50
Lens of the eye per year	300	30
For the skin per year (if this is the result of radioactive contamination the limit shall apply to the dose averaged over any area of 100 cm²)	500	50
Hands, forearms, feet or ankles per year	500	50

In the case of apprentices and students the following conditions apply:

1. The dose limits for apprentices and students aged 18 years or over who are training for employment involving exposure to ionizing radiation or who, in the course of their studies, are obliged to use sources shall be equal to the dose limits for exposed workers.
2. The dose limits for apprentices and students aged between 16 and 18 years who are training for employment involving exposure to ionizing radiation or who, in the course of their studies, are obliged to use sources,

shall be equal to three-tenths of the annual dose limits for exposed workers.

3. The dose limits for apprentices and students aged 16 years or over who are not subject to the provisions of paragraphs 1 and 2 and for apprentices and students under 16 years shall be the same as the dose limits for members of the public. However, the contribution to the annual doses that they are liable to receive by virtue of their training shall not exceed one-tenth of the dose limits specified for members of the public and the dose during each single exposure shall not exceed one-hundredth of those dose limits.

Classification of Areas

The Council Directive referred to above includes a statement on classification of areas.

In working areas where the exposure is not liable to exceed one-tenth of the annual dose for exposed workers, it is not necessary to make special arrangements for the purposes of radiation protection. In working areas where the annual exposure is liable to exceed one-tenth of the limits of annual dose laid down for exposed workers, the arrangements must be appropriate to the nature of the installation and sources and to the magnitude and nature of the hazards. The scope of the precautions and monitoring, as well as their type and quality, must be appropriate to the hazards associated with the work involving exposure to ionizing radiation.

A distinction shall be made between:

i. controlled areas which are areas in which the doses are likely to exceed three-tenths of the limits of annual dose laid down for exposed workers;
ii. supervised areas which are areas not considered as controlled areas and in which doses are liable to exceed one-tenth of the annual dose limits laid down for exposed workers.

Controlled and supervised areas

i. Controlled areas must be demarcated and control of access by appropriate warning signs is essential in both controlled and supervised areas.
ii. Depending on the nature and extent of the radiation hazards, the following are required:
    a) radiological environmental surveillance shall be organized in controlled and supervised areas and, in particular activities, doses and dose rates as the case may be shall be monitored and results recorded;
    b) in controlled and supervised areas, working instructions appropriate to the radiation hazard shall be laid down;

c) the hazards inherent to the sources shall be indicated in controlled areas;
d) signs indicating sources shall be displayed in controlled and supervised areas.

Qualified experts shall be engaged in the discharge of these duties. The minimum requirement for a controlled area shall be the control of access by warning signs.

Classification of Exposed Workers

Exposed workers should be classified as follows:

Category A—those who are likely to receive a dose greater than three-tenths of a limit of annual dose.

Category B—those who are not liable to receive this dose.

Exposed workers must be informed of the health risks in their work, the precautions to be taken, and the importance of complying with the technical and medical requirements. Apprentices and students who are training for employment involving exposure to ionizing radiation or who in the course of their studies are obliged to use sources must be given adequate training and appropriate information regarding the risks in their work.

Examination and Testing of Protective Devices and Monitoring Instruments

These functions must be carried out by qualified experts and must include the following from the point of view of radiation protection:

a) prior critical examination of plans;
b) acceptance of new installations;
c) regular checking of protective devices and techniques;
d) regular checking that measuring instruments are serviceable and correctly used.

Assessment of Exposure

Measurements of dose rates, fluence rates (the number of particles entering a sphere of given cross-sectional area in a stated time) and atmospheric concentration must be carried out indicating the nature and quantity of the radiation.

Individual Monitoring

Systematic assessment of individual areas is essential for Category A workers. In the case of accidental or emergency exposure the absorbed dose shall be assessed whether whole or partial body irradiation has occurred.

Medical Surveillance

A medical practitioner approved by the competent authorities must be appointed for surveillance of Category A workers. He is responsible for interpreting the results of individual doses in relation to the health of the workers. Results of the collective monitoring measurements, exposure records, and accident or emergency exposure must be kept for a period of at least 30 years.

The medical surveillance of exposed workers of Category A must include:

a) a pre-employment medical examination;
b) general medical surveillance;
c) periodic reviews of health;

and may possibly continue after cessation of work. Special surveillance must be provided in each case where the dose limits given in the section on Permissible Doses are exceeded.

Protection of the Population

It is unlikely that the general laboratory worker will ever be in such a position that the results of an accidental mishap in his work would cause radiation hazard to the general population. Nevertheless, in a few isolated instances it could happen and he should be aware of his responsibilities if such an event takes place. A summary of the preventative measures from the Council Directive is as follows:

a) The work must be subjected to rigorous examination to ensure complete operational protection. This includes examination of proposed installations, measuring instruments, emergency plans, waste discharge and the effectiveness of technical protective devices.
b) Health surveillance of the population.

Implementation of a Control System in the Laboratory

The scheme of operation described in this section for safe working in laboratories where there is a radiation hazard is broadly based on the publication "Guidance Notes for the Protection of Persons Exposed to Ionizing Radiation in Research and Teaching" (London, HMSO, 1976) and on experience gained in the authors' own laboratories.

A Controlling Authority must be established which has ultimate responsibility for safe working in the establishment. This authority is responsible for drawing up a code of practice and delegation of responsibility. The latter and all communications in radiation work must be communicated in writing and their receipt similarly acknowledged.

The Controlling Authority is responsible for appointing a sufficient number of trained Radiological Safety Officers to carry out the duties specified in the Code. The duties of the Radiological Safety Officer comprise:

i. designation of Responsible Persons;
ii. supervision of health, dose, material, instrument and source testing records;
iii. approval of procedures for installation, use and maintenance of radioactive devices, materials, and X-ray instruments;
iv. ensuring compliance with local rules and current regulations,
v. investigation of emergencies.
Any duty of a Responsible Person where appropriate.

The Controlling Authority must decide which persons come within the Code and which of these are designated persons, i.e. work within controlled or supervised areas (for definition see pp. 102–103).

It is noted however that persons following an approved scheme in controlled areas are not designated, an approved scheme of work being defined as one in which the Controlling Authority is satisfied that a person is unlikely to receive more than three-tenths of annual maximum permissible doses given on pp. 100–101, i.e. Category B workers. The Certificate of Registration must be displayed in a prominent position. The Controlling Authority must arrange for the following:

a) All controlled areas to be identified and suitably marked.
b) Records to be kept of radiation doses, transfer of records, cases of contaminated skin, hair, clothing, etc., results of tests of, and by, monitoring equipment, staff health records, sealed sources stock, unsealed sources stock, leakage tests, and investigations into emergencies.
c) Radiation emitted by sealed sources, instruments, and unsealed radioactive substances to be reduced to the lowest possible level to minimize external radiation received by the body.
d) Measures to be taken to minimize deposition within the body of radioactive substances by means such as inhalation, ingestion, etc.
e) Monitoring of the working environment—this is carried out by a Responsible Person, appointed by the Controlling Authority, who is charged with carrying out special assignments for the Authority which, in the authors' laboratories, include:
i. designation of radiation workers,

ii. instruction of designated radiation workers,
iii. notification of changes in the list of radiation workers,
iv. preparation of procedures for installation, use and maintenance of sealed and unsealed sources and X-ray apparatus for approval by the Radiological Safety Officer.

f) Personal monitoring to be carried out using methods which are relevant to the nature of the hazard. The usual forms of personal monitoring are by photographic film badge, thermoluminescent indicator, or ionization chamber. These are normally worn on the chest but they may be positioned elsewhere, e.g. the finger or wrist, if required. The thermoluminescent type of monitor gives a more accurate assessment than the film badge over long periods of wear and it enables dose record-keeping to be carried out automatically. The ionization chamber gives an instantaneous measurement of the radiation received.

g) Monitoring instruments and facilities for their use to be provided. They must be tested and calibrated by a suitably qualified person when they are first taken into use and retested and recalibrated at least annually thereafter. Where possible the operator must be able to make a quick and simple check of the instrument's performance. Details of typical monitoring instruments are given later.

h) Medical supervision by a qualified Medical Adviser. An initial medical examination of designated persons must be carried out within 14 months prior to starting work. A blood examination must be included consisting of, or including:

i. in the case of the red blood cells, a measurement of the packed cell volume;
ii. in the case of white blood cells, an estimate of the number present per cubic millimetre of whole blood;
iii. a search for abnormal cells and a description of any seen;
iv. an estimate of the haemoglobin in grams per 100 millilitres of whole blood.

Designated persons must be re-examined annually to check their fitness for work unless monitoring shows that they are receiving less than the dose limit for designated persons. Records must be kept by the Medical Adviser of medical examinations and they must be safeguarded as confidential documents.

Special Precautions Regarding Laboratory Equipment

These precautions apply to X-ray installations and other apparatus in which ionizing radiations are used. The installation of such equipment must be supervised by the Radiological Protection Officer and it must preferably be shielded so that all radiation is contained. If this is not possible then the area must be clearly marked. It should be possible to shut off the ionizing radiation quickly and to evacuate the area rapidly. When the apparatus is about to be energized, either a light or audible signal should warn persons in the vicinity, and one or the other should continue to operate while the apparatus is energized. Whenever apparatus is moved or modified or the shielding changed it must be re-monitored.

X-ray instruments

While it is true to say that early X-ray instruments were not too well guarded, it is equally true to report now that present-day instruments are

PERMIT TO WORK INVOLVING A HAZARD FROM IONISING RADIATIONS

№ 199

Room No.	Valid from hrs / /19
	Valid to hrs / /19

If the work is not completed in the time allotted above, a new certificate must be issued.

Item of equipment and description of work..

....

....

....

Special requirements ....

X-ray generator isolated. Special protective clothing to be worn.

Approved scheme of work to be followed.

(No more than two of these items to be deleted if applicable.)

If the X-ray generator is isolated the key must be handed to the Responsible Person or nominee.

Additional Instructions ....

....

....

....

The Responsible Person or his nominee must be informed as soon as the work is complete or this certificate expires (whichever is sooner).

H 6432 78

The above requirements have been fulfilled and it is safe to proceed with the above work.	Date.................... Time.................... Signed .... (Responsible Person or nominee)
I have read and understood this certificate and undertake to work in accordance with the conditions above.	Date.................... Time.................... Signed .... (Permit Holder)

CERTIFICATE CLOSURE

Work on the job detailed above has finished and is complete /incomplete. All guards and protective devices have/have not been replaced.	Date.................... Time.................... Signed .... (Permit Holder)
Part A or B must be completed A The generator has been locked OFF	Date.................... Time.................... Signed .... (Responsible Person or nominee)
B Monitoring has been carried out, the radiation level in the immediate vicinity is less than 0.75 millirads/hr and the equipment is now safe to operate.	Date.................... Time.................... Signed .... (Responsible Person or nominee)

Fig. 6.3. An example of a permit to work.

well equipped with safety devices. Interlocks are fitted so that the X-ray tube cannot be energized unless safety devices are in position and shielding is now more efficient. The main problems occur during setting up of the instrument and alignment of the beam when it is necessary to remove shielding. This work must be confined to a small number of highly experienced persons and it may be necessary for them to wear protective clothing while carrying out the work. Experience has shown that this objective is best achieved where operation as well as maintenance of the instruments is restricted to a small number of named personnel and persons not listed are not permitted to use the instruments. A permit to work system is also in operation and a copy of this is shown in Fig. 6.3. It is an essential requirement of this scheme that the X-ray generators are fitted with isolation switches which can be locked in the "off" position. It is impossible for the generators to be switched on without inserting the key and they cannot be switched off unless the key is withdrawn. In the case of diffraction spectrometers, where samples have to be inserted by hand, mechanical interlocks are used so that opening of the sample port automatically places a barrier across the X-ray beam.

Particle accelerators

This section includes Van de Graaff electrostatic generators, neutron generators, linear accelerators, cyclotrons, betatrons, synchrocyclotrons, synchrotrons and proton synchrotrons. As the safety problems associated with these instruments are complex, only a brief outline of general precautions will be given here.

The instruments must be adequately shielded or installed in a special room, the door of which is fitted with an interlock so that the room cannot be entered with the equipment switched on. The machine and the neighbouring area must be carefully monitored and this should be repeated for every change in operating conditions as the radiation can be complex. Induced radioactivity may also be present.

These instruments, as well as X-ray spectrometers, present electrical safety problems besides mechanical and chemical hazards and appropriate precautions have to be taken against them.

Sealed sources

Sealed sources are usually delivered by the supplier in containers which must be of adequate mechanical strength and design to prevent accidental leakage of radiation. All sources and/or containers must be marked with the standard trefoil (in BS 3510 1968: "A basic symbol to denote the actual or potential presence of ionizing radiation") and each source must be clearly marked with the source number or other mark identifying nature and activity. They should be returned to their containers when not in use. If a sealed source is likely to be used in several locations a daily log of its use must be kept.

If a source is lost or damaged in any way, the appropriate authorities (in the UK, the Police and Radiological Protection Board) and Responsible Person must be informed immediately and all unauthorized persons excluded from the area. A record must be kept, listing full details of all sealed sources and retained until at least two years after disposing of them.

The method of working and the installation of the sealed source must be designed to minimize exposure to radiation. Every source must be tested for leakage or surface contamination at least once every two years and the results recorded. Any leakage must be reported to the Responsible Person.

Unsealed radioactive substances

In working with unsealed sources special precautions against hazards due to internal as well as external radiation are necessary. Special clearly marked areas must be set aside for working with unsealed sources; in the case of work with very small amounts of radioactive traces (not alpha-emitters) and subject to the approval of the Radiological Protection Officer, it is not necessary to work in a controlled area. Warning notices must be displayed at the boundaries of the area while the work is in progress.

Particular attention must be paid to the design of the laboratory. Surfaces should be smooth and free from cracks; areas such as ledges that can collect dust must be eliminated. Ideally, operations should be carried out in total enclosure such as a glove box, although for less hazardous materials a fume cupboard will suffice.

Facilities must be available for changing into and out of protective clothing. This should be monitored regularly and cleaned under special circumstances if the activity gets to a high level. In cases of gross contamination it should be treated as radioactive waste. Further details of procedures are contained in the Guidance Notes[2] identified in the reference section at the end of this chapter.

The container must be monitored on receipt and handling precautions adopted according to the level of the radiation, on the recommendation of the Responsible Person. Working areas and equipment must be monitored at the end of each period of use and a record kept of the results of the checks for two years after the last entry. Apparatus must be carefully examined to minimize risk of breakage or spillage and care should be taken to prevent contact of the source with parts of the body; mouth suction must not be used with pipettes. To prevent contamination of benches etc., absorbent paper may be placed on its surface and disposed of at the end of the work. Such waste must be checked for radioactivity before disposal. Eating, drinking, taking snuff, smoking, and using cosmetics are prohibited, and particular care should be taken when changing clothes on leaving the work. Cuts and other breaks in the skin particularly on the hands and forearms must be carefully covered before work.

When not needed in the laboratory radioactive substances must be stored in a secure place near to the work area to reduce possible exposure during transit. The store must be secure from fire and flooding and marked with the radioactive trefoil. The advice of the local Fire Brigade Officer should be sought when positioning the safe. The contents of the safe must be recorded by the Radiological Safety Officer.

In the United Kingdom all laboratories working with radioactive

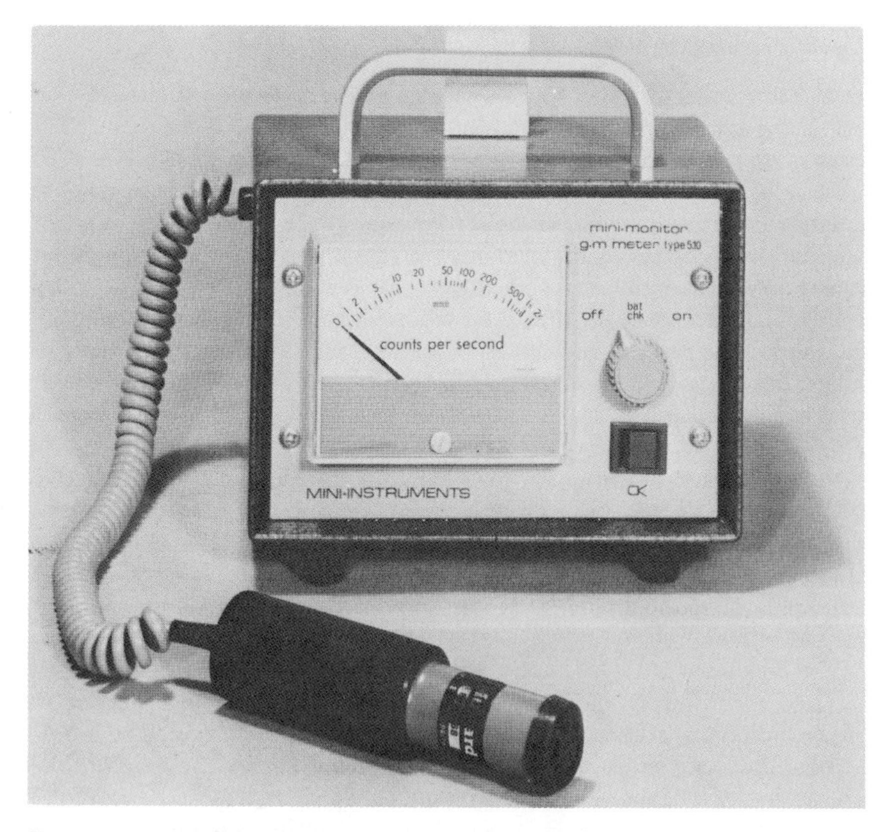

Fig. 6.4. A typical radiation monitor. When used with different probes it can measure radiations due to β emitters, β-α emitters, X- and α-radiation. (Photograph courtesy Mini Instruments Limited).

materials must be registered with the Department of the Environment and authorization from the department is necessary before starting work and before disposing of such waste. Disposal must be supervised by the Responsible Person and the Radiological Protection Officer notified.

Monitoring Instruments

The modern monitor, of which there are numerous examples on the market, is a small compact battery-operated instrument using either a Geiger tube, ionization chamber or scintillation detector system. Monitors are available for X, alpha, beta and gamma radiation and neutron detection. Indication is by meter or an earphone can be used to give an audible signal. Typical of these instruments is the Mini-monitor[3] shown in Fig. 6.4. The basic instrument can be used in combination with a number of probes for the detection of β emitters, β–α emitters, X-radiation and α-radiation of intensity above $37\,\mathrm{Bq\,cm^{-2}}$. Instruments must be calibrated and checked every 14 months for the energy levels likely to be encountered.

Emergency Procedure

Emergency procedures must be drawn up and periodically tested by practice exercises to deal with any unplanned excessive exposure to ionizing radiations. Pre-planning must ensure that all necessary safety equipment is available for personal protection, isolation of the area, cleaning spillage etc., and that external safety services such as medical, ambulance, fire, police have been informed that radioactive materials are being used. A typical emergency procedure would be as follows:

1. Any case of suspected exposure, or loss of radioactive materials must be reported immediately to the appropriate Responsible Person and/or the Radiological Safety Officer who will call on medical advice/help if necessary.
2. The contaminated high dose rate area must be isolated/evacuated immediately. No attempt may be made to decontaminate a contaminated area without the approval of the appropriate Responsible Person or the Radiological Safety Officer.

   The original working procedure must specify a decontamination procedure for emergency use.
3. The Radiological Safety Officer will notify the Controlling Authority and other relevant persons.
4. Reduction of radiation hazard. If a very high rate is expected the original working procedure must include a scheme for ensuring that no person dealing with any emergency receives a whole body dose in excess of 0.12 sievert.

   If a sealed source is involved, the source should be retrieved by detection or the hazard reduced by containerization or shielding.

   If unsealed, attempts must be made to limit the dispersal of the substance, or, if a gas or vapour, to disperse them as quickly as possible.

If the emergency arises from an X-ray machine, the machine should be switched off —if not at the machine itself then by use of mains isolators or water supply switches.

5. Contaminated persons should not proceed into inactive areas until they have been monitored and/or decontaminated. Serious injuries must be dealt with first, however.
6. If an accident occurs involving loss or spillage of radioactive material and outside help is required, the following services are available:

Addresses of organizations able to be of assistance should be included here.

After the incident

1. Medical surveillance of persons must take account of any residual contamination.
2. If any person has received a dose greater than the maximum permitted levels, a full investigation must be made by the Radiological Safety Officer and a full medical examination carried out.
3. Affected areas must be fully decontaminated and equipment tested before re-use; access to these areas must be restricted to properly trained and equipped persons.
4. The Radiological Safety Officer must carry out a full investigation of the emergency and report in writing to the Controlling Authority.

Disposal of Radioactive Waste

In the UK all establishments keeping and using radioactive materials (subject to certain exceptions) are required to register with the appropriate Government body and an authorization must be obtained for the accumulation and disposal of radioactive waste. Disposal of such material must therefore be done under the supervision of the Radiological Safety Officer or Responsible Person who is responsible for obtaining the necessary authorization and detailing the method to be followed. Material of short half-life can be stored until the activity has reached a safe level; highly active materials may have to be disposed of by burial at sea in special containers.

ULTRAVIOLET RADIATION

Equipment which contains sources of ultraviolet radiation is widely used in laboratories. Exposure of the unprotected eyes and skin to the radiation can lead to temporary inflammation, the severity depending upon the duration and intensity of the exposure. Longer term effects may result from repeated exposure. Details of the hazards arising, and the protective measures required are summarized in the following sections.

Nature of Radiation

Ultraviolet radiation ranges in wavelengths from 400 nanometres (nm) to 10 nm. The hazard to health is from wavelengths longer than 200 nm.

Wavelengths between 315 nm and 200 nm are the most harmful, and the hazard rises steeply between these limits to a maximum at 270 nm. The source may radiate discrete wavelengths, or a continuous spectrum, dependent upon how the radiation is produced.

Effects

Exposure of the unprotected skin may give rise to erythema (similar to sunburn). Exposure of unprotected eyes may give rise to keratitis (inflammation of the cornea) and conjunctivitis. The severity of the effects depends upon the time of exposure and the wavelength, and the intensity of the radiation. Guidance on maximum permitted times of exposure without protection (MPE) is given in a booklet[4] which is summarized below (permission to reproduce information from this source is acknowledged).

The Maximum Permissible Exposures (MPE) are expressed in radiant exposure units of joules per square metre ($J\,m^{-2}$), the total ultraviolet radiation energy falling on $1\,m^2$ of surface, or in irradiance units of watts per square metre ($W\,m^{-2}$), the average radiant exposure per second over the exposure period.

Table 6.2. Relative spectral effectiveness ($S\lambda$).

Wavelength	*MPE*	$S\lambda$
(nm)	($J\,m^{-2}$)	
200	1000	0.03
210	400	0.075
220	250	0.12
230	160	0.19
240	100	0.30
250	70	0.43
254*	60	0.5
260	46	0.65
270	30	1.00
280	34	0.88
290	47	0.64
300	100	0.3
305	500	0.06
310	2000	0.015
315	10000	0.003

*Mercury lamp—resonance emission line.

Wavelength range 400–315 nm

i. Total irradiance on unprotected eyes and skin for periods of greater than 1000 s should not exceed 10 W m^{-2}.
ii. Total radiant exposure on unprotected eyes and skin for periods of less than 1000 s should not exceed 10^4 J m^{-2}.

Wavelength range 315–200 nm

The radiant exposure on the unprotected eyes and skin should not exceed, within any 8-hour period, the values given in the Table 6.2.

As the values given in the above Table 6.2 apply only to sources emitting essentially monochromatic ultraviolet radiation, a calculation must be made to assess the effective irradiance of a broad-band source. This is carried out as follows. The MPE for a broad-band source is calculated by summing the relative contributions from all its spectral components, each contribution being weighted by the relative spectral effectiveness, Sλ. Values of Sλ for the range, 200–315 nm, are also given in Table 6.2.

$$E_{eff} = \Sigma\, E\lambda . S\lambda . \Delta\lambda$$

where E_{eff} = effective irradiance relative to monochromatic wavelength 270 nm (W m^{-2}).

Eλ = spectral irradiance at wavelength (W m^{-2} nm^{-1})
Sλ = relative spectral effectiveness
Δλ = band width employed in the measurement or calculation of Eλ (nm)

The maximum permissible exposure, expressed in seconds, may be calculated by dividing the MPE for 270 nm radiation (30 J m^{-2}) by E_{eff} (W m^{-2}).

Table 6.3. Maximum permissible exposure in an 8-h period.

E_{eff} (W m^{-2})	Maximum permissible exposure
10^{-3}	8 h
8×10^{-3}	1 h
5×10^{-3}	10 min
5×10^{-1}	1 min
3	10 s
30	1 s
3×10^2	0.1 s

Data on intensity of radiation from various sources are given in several publications.[5,6]

Values of maximum permissible exposure are given in Table 6.3.

Protection

Where sources are powerful enough to be a hazard, protection against over-exposure may be achieved by a combination of:

a) administrative control measures;
b) engineering control measures;
c) personal protection measures.

Emphasis should be placed on administrative and engineering control measures, to minimize the need for personal protection.

Administrative control measures

All users of equipment must be made aware of the nature of the hazards involved. Suitable warning signs must be displayed on or near the equipment. Other measures include limitation of access, distance of the user from the source, and limitation of exposure time.

Engineering control measures

Indiscriminate emission of ultraviolet radiation into the workplace is not allowed, and can be prevented by carrying out the process in a sealed housing, or by providing a screened area. The intensity of reflected radiation within a screened area may be reduced by the use of matt black paint. Where the emission from a source is above the MPE at the normal working distance, screens should be fitted with power interlocks.

Personal protection

Where the nature of the work requires a user to work close to an unscreened ultraviolet source, personal protection must be used. The eyes must be protected by goggles, spectacles or a face shield which absorb ultraviolet waves. Consideration must also be given to the need for protection of the hands (cotton gloves) and the skin of the face (a face shield).

INFRA-RED RADIATION

Infra-red radiation is readily absorbed by surface tissues to produce a heating effect and it does not cause any deep injuries. The heat is not readily dispersed by the lens of the eye and a cataract may be produced. Sources emitting infra-red radiation should be shielded by heat-absorbing screens and eye shields of an approved type should be worn to protect the eyes.

LASER BEAMS

General

The name "laser" is derived from the term Light Amplification by Stimulated Emission of Radiation. The function of a laser is to produce a beam of light of high intensity which has been used in, for example, surgery, surveying, and communications. It is now finding increasing use in research laboratories in the context of fluid flow, chemical kinetics, and light-scattering experiments.

Lasers are manufactured in a variety of forms including units with continuous and pulsed outputs of both visible and invisible radiation. All units can, if sufficiently powerful, damage the skin and in particular the eyes if exposure to the radiation occurs. The high-power pulsed unit is particularly hazardous as the short pulse of light can cause widespread damage around the point at which the beam contacts the tissue. Continuous wave lasers can also cause damage to the eyes and skin but this tends to be localized to the point of contact of the beam on the eye or skin tissue. Lasers emitting visible and infra-red radiation will present a hazard to the retina of the eye in the form of burns, those emitting invisible radiation of ultraviolet and far infra-red represent a corneal hazard where exposure can also cause the formation of cataracts. In general, the skin is much less sensitive to laser light. In the case of visible and infra-red units very large power lasers may result in a form of sunburn or blisters. Other problems may be encountered with powerful lasers where the skin is sensitized by repeated local exposure; examples of this are, however, extremely rare.

In the context of safety, lasers can be considered in the following two groups.

Group A: Intrinsically safe units and those falling into classes I and II as defined in the British Standard Guide to the Protection of Personnel Against Hazards from Laser Radiation (1972 – under revision) BS 4803 and the American Standard, The Safe Use of Lasers, ANSI Z136.1-1976.

Group B: All pulsed lasers and continuous wave lasers whose power is greater than 1 mW as defined in classes IIIa, IIIb and IV of BS 4803 and ANSI Z136.1-1976.

The eye can be damaged by 50μW of light entering the pupil so suitable precautions should always be taken to minimize the chance of exposure.

As a guide to the magnitude of the risks associated with particular laser units BS 4803 recommends that:

1. With class IIIa lasers (< 5 mW) precautions are only required to prevent direct viewing of the beam.
2. In the case of class IIIb units (5 mW–500 mW continuous output) direct viewing and specular reflections viewed by the unprotected eye are hazardous.
3. With class IV units (continuous powers greater than 500 mW) direct viewing, secondary and diffuse reflections can be dangerous and great care must always be taken.

Skin damage is possible with classes IIIb and IV and precautions should be taken accordingly.

In addition to the hazard associated with the main beam there are the hazards due to the electrical source units used to power the laser.

The following code of practice covering the use of lasers has been drawn up for use in the authors' laboratories.

Personnel

1. Only authorized personnel are allowed into a laboratory using lasers.
2. This authorization must be obtained from the manager in charge of the work in the laboratory and from the Laboratory Safety Officer.
3. Authorization must be given in written form.
4. Prior to authorization to work with lasers, all workers must have an ophthalmological examination with special emphasis on the condition of their retinas.
5. A list of authorized workers will be kept by the manager in charge of work involving lasers. All persons on the list will have an ophthalmological examination at six-monthly intervals. Persons working on class IIIb and IV lasers will also have an inspection for skin damage at the same time.
6. Persons who stop working with lasers and therefore do not require medical examination, will be removed from the list of authorized staff.
7. Prior to authorization the manager in charge of the work will ensure that staff are properly trained in safe working procedures and know the potential hazards of lasers and associated equipment. They must read documents BS 4802 and ABSI Z136.1—1976.
8. Any accidental exposure of the eye must be reported to the medical department.
9. Eye protection should be worn as general protection whenever practical when using Group B lasers BUT NEVER USED FOR DIRECT VIEWING.
10. All personnel visiting a laser laboratory outside the company should consult the Laboratory Safety Officer.

Work Area

1. Laser equipment should be set up away from other work in a close controlled and screened room or area.

2. The room should be well lit to avoid enlarging the pupil of the eye thereby increasing the possibility of damage.
3. Reflected laser beams can be as dangerous as the direct beam. The area should therefore be free from polished and reflective surfaces.
4. The control area should be periodically monitored for any toxic gases which may be generated during the work, for example, ozone etc.
5. Suitable warning signs must be placed at all entrances to rooms where lasers are used. These signs should be illuminated and interlocked to the laser power supply. When actuated the signs must prohibit access to all but authorized personnel.
6. Where necessary, blinds and screens must prevent the possibility of radiation straying out of the laser control area.
7. In the case of infra-red beams, suitable non-flammable materials must be used to provide physical barriers around the laser area. These lasers should not be used in the vicinity of flammable or explosive materials.

Equipment

1. Energy sources for lasers are essentially high-voltage equipment and standard safety precautions for dealing with such equipment must be followed.
2. The laser beam lenses, filters, and other optical components must be shielded to prevent accidental deflection or reflection of the laser beam.
3. All equipment operating in excess of 15 kV must be surveyed for possible emission of X-radiation.
4. Cryogens, such as liquid nitrogen, for reducing the temperature of laser equipment, should not be handled without proper hand, face, and body protection.
5. Maintenance work must only be carried out when the power is switched off.
6. All lasers must have warning labels with an appropriate cautionary statement fixed to a conspicuous place on the laser housing or control panel. This label should include the class of laser as defined in BS 4802 and ANSI Z136.1—1976. The end of the laser from which the radiation emerges should be clearly labelled with the type of radiation, invisible or visible, stated.
7. When not in use all lasers of Group B should be rendered unusable by removal of a key control. Where a key control is not built into the system the power supply should be isolated using a lockable switch. Keys should be in the custody of the authorized personnel responsible for laser work in each laboratory.
8. Eye protectors must be of an approved type (British Standard 4899) and must be checked regularly for damage. They must be chosen to match the wavelength of light emitted by the laser in use.
9. Any laser brought into the company's premises must be reported to the Laboratory Safety Officer.
10. Target screens must be of diffuse absorbent materials to prevent specular reflection.

Operation

1. A "countdown" procedure should be followed before energizing the beam. All personnel should look away and close their eyelids before the firing signal to avoid exposure to any possible reflections. They must also be behind the laser beam source and well to the right or left of the proposed beam path before firing.
2. Only the minimum number of people required to operate the laser should be in the

control area during operation. In the case of class IV lasers, however, at least two people should always be present during operation.

3. Only authorized personnel should be allowed to set up, adjust and/or operate laser equipment.
4. Equipment should only be operated for the minimum time required for the specific experiment.
5. Laser equipment should not be left running unattended.
6. Eye protectors must be worn as general protection whenever there is a possible risk of eye damage.
7. All laser equipment should be assembled at or below the operators' chest height.
8. Particular care should be taken if operating invisible radiation laser units. The radiation from IR and UV lasers can cause skin complaints and cataracts, and full shielding of the beam path is essential. Infra-red units are generally very powerful and can cause severe skin burns and eye damage. Special attention should be given to the possibility of the radiation forming skin sensitizing agents. Eye protection must always be worn.
9. Pulsed lasers represent an increased hazard to the operator. Helium Neon low-power lasers (class II) should be used as an alignment guide but not relied on as an indication of where the beam will go. Eye protection must always be worn with these systems as exposure of the eye to intense pulsed radiation can cause widespread damage to the retina.

Exceptions

In exceptional circumstances, to be determined by line management in conjunction with the Laboratory Safety Officer, some of the provisions above may become superfluous as a result of more stringent controls applied to limit the emission from equipment of harmful radiation.

Each such case will be considered on its own merits, the necessary safeguards being agreed between line management and the Laboratory Safety Officer and communicated to affected parties in writing prior to the commencement of work.

MICROWAVE AND RADIO-FREQUENCY EQUIPMENT

The microwave radiation band generally includes all electromagnetic radiation in the wavelength range of 1 cm to 300 cm (30 GHz to 100 MHz) and the radio-frequency band covers the microwave range and extends to a wavelength of 30 Km (10 KHz).

There is a double hazard with this type of equipment both from the normal electrical hazards and from local burns arising from contact with the work coil or from heat produced by induced currents in such conductors as rings and watches. All radio-frequency heating work coils must have a substantial earth connection and be isolated from the high voltage by blocking condensers of adequate working voltage. The work should only be carried out by authorized personnel who are fully acquainted with the potential hazards.

Areas in which microwave power density of more than $0.01\ \mathrm{W\,cm^{-2}}$ is detected or suspected should be considered hazardous and posted with warning signs. Personnel exposure to lower levels should be held to a minimum.

A draft directive is being prepared by the European Commission which will limit exposure to microwave radiation in the range 300 MHz to 30 GHz to $100\ \mathrm{W\,cm^{-2}}$ as averaged over any period of 0.1 h.

The energy beam must not be directed towards personnel. The probable directions of reflected beams should also be considered.

Microwaves over $3000\ \mathrm{MHz\,s^{-1}}$ are usually reflected or absorbed by skin and can be felt through heating of surface tissue. From 1000 to $3000\ \mathrm{MHz\,s^{-1}}$ penetration of the skin and fat layer occurs while below $1000\ \mathrm{MHz\,s^{-1}}$ pene-

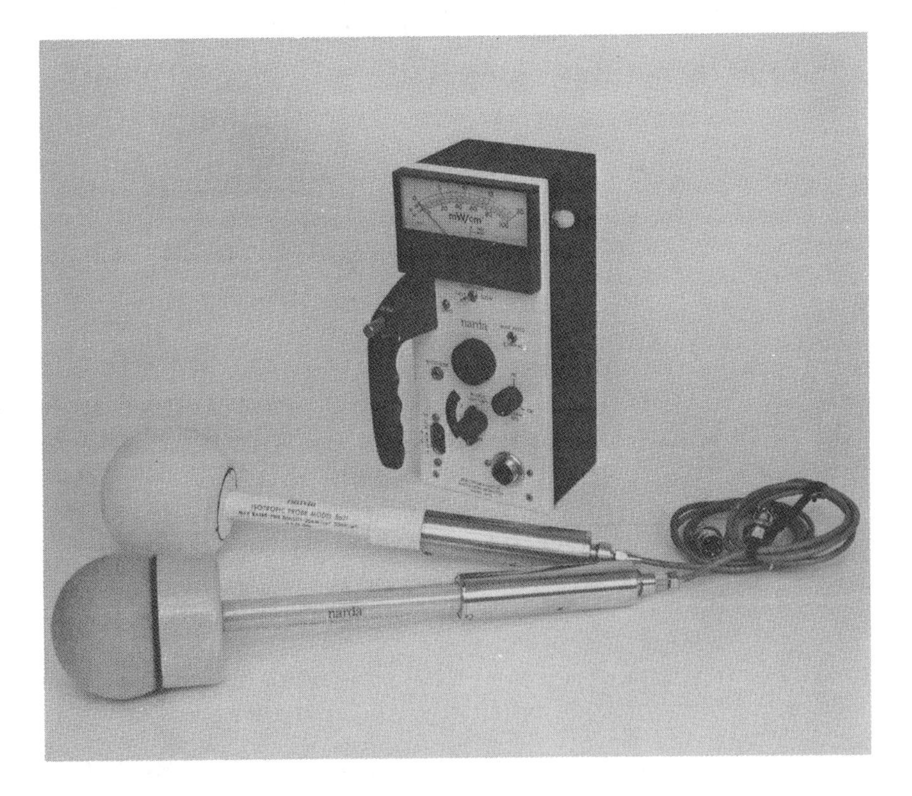

Fig. 6.5. A typical instrument for monitoring microwave radiation. (Photograph courtesy The Narda Microwave Corporation).

tration of the deep tissues takes place without subjective awareness of heating. Eyes and those organs which cannot readily dissipate heat are most vulnerable.

Rings, watches, and similar metallic objects must not be worn nearby. Persons who have had bones pinned with metal should be exceptionally careful not to expose themselves to this type of energy.

Suitable screens must be provided to avoid dangerous voltages being induced in neighbouring metallic equipment, which should be effectively earthed.

Where microwave generating equipment such as klystrons or magnetrons operate, spurious X-rays are generally present.

In addition to radio-frequency hazards be alert to high dc potentials on components.

Guard against sharp edges or points which can emit corona discharges and cause burns.

If anything abnormal occurs while an experiment is in progress, the heaters should be switched off immediately.

Adjustments or modifications to the equipment may only be carried out by authorized personnel. Any such modifications must only be carried out after isolation of the equipment from the mains supply to prevent accidental switching on while the work is in progress.

Microwaves can ignite flammable solvents, photoflash bulbs, and explosive fuses.

Microwave Radiation Monitors

Monitoring equipment is available for the detection and measurement of microwave equipment such as ovens, heaters, driers, and medical equipment. A typical instrument is marketed by the Narda Microwave Corporation of Plainview, New York and available in the UK through their agents Aveley Electrics Ltd. of Chessington, Surrey (see Fig. 6.5). The instrument covers the range of frequencies from 10 MHz to 26 GHz and power reading ranges from 0.2 mW cm^{-2}. It has an extended probe and connecting cable so that the operator can monitor equipment from a safe distance. Simpler and cheaper devices are available which only indicate the relative power of the microwave radiation.

REFERENCES

1. Council Directive, 15 July 1980 (80/836 EURATOM), Official Journal of the European Communities, 17 Sept. 1980 (L246, Vol. 23).

2. Guidance Notes for the Protection of Persons Exposed to Ionising Radiations in Research and Teaching (1968), HMSO, London.
3. Mini-instruments Ltd, of Burnham on Crouch, Essex, UK.
4. Protection against Ultraviolet Radiation in the Workplace, National Radiological Protection Board, Harwell, Didcot, Oxon. OX11 0RQ, UK.
5. Hughes, D. (1979). Hazards of Occupational Exposure to Ultraviolet Radiation. Occupational Hygiene Monograph No. 1, University of Leeds Industrial Services Ltd, UK.
6. Koller, L. R. (1952). "Ultraviolet Radiation", John Wiley & Sons, London and New York.

FURTHER READING

Boursnell, J. C. (1958). "Safety Techniques for Radioactive Tracers", Cambridge University Press.

"Code of Practice against Radiation Hazards" 6th edn. (1973). Imperial College, London SW7 2AZ.

Everett, K. (1966). "University of Leeds Safety Handbook", University of Leeds Safety Committee.

"A Guide to the Safe Use of X-ray Crystallographic and Spectrometric Equipment" (1977). Association of University Radiation Protection Officers, Monograph Series No. 1, produced by P. J. F. Griffiths, University of Wales Institute of Science and Technology, Cardiff.

Hamilton, M. (1979). "Microwaves—No Cause for Hysteria", *Occupational Safety and Health*, **9,** No. 12, 14.

"Health and Safety in Welding", 2nd edn. (1965). The Institute of Welding, 54 Princess Gate, Exhibition Road, London.

"Ionising Radiations, Supplementary Proposals for Provisions on Radiological Protection and Draft Advice from the National Radiological Protection Board to the Health and Safety Commission" (1979). Health and Safety Commission, HMSO, London.

"Ionising Radiations, Proposals for Provisions on Radiological Protection" (1979). Health and Safety Commission, HMSO, London.

"The Ionising Radiations (Unsealed Radioactive Substances) Regulations", 1968. The Ionising Radiations (Sealed Sources) Regulations, 1969. Radioactive Substances Act, 1960. HMSO, London.

"Living with Radiation" (1973). National Radiological Protection Board, HMSO, London.

"Radiological Protection in the Universities" (1966). The Association of Commonwealth Universities, 36 Gordon Square, London, WC1.

"Recommendations of the International Commission of Radiological Protection" (1965). Pergamon Press, Oxford.

'Safety Handbook", University of Durham Science Site and Observatory.

"Safety Precautions in the Use of Electrical Equipment" (1968). Imperial College of Science and Technology, London.

Steere, N. V. (1971). "Handbook of Laboratory Safety", C.R.C. Press Inc., Boca Raton, Florida, USA.

7

Compressed Gases

The handling of compressed air and gases in laboratories is not always treated with the seriousness the hazard potential demands. A walk round a laboratory complex after normal work has ceased will prove illuminating in this context: cylinder valves are not properly closed; there are gauges for gases other than those actually in cylinders; others which should be upright lie in a horizontal position; others upright but not fixed are amongst the defects probably found. The question of whether the cylinders should be inside at all, or stored outside, will be discussed later. The use of compressed air for blowing down clothes after work amongst fine dusts and powders is also common. There are recorded cases of limbs blown up to elephantine size when compressed air had been inadvertently applied to a minute skin puncture. In an automobile manufacturer's assembly plant a booklet was issued to all employees showing by illustrations the unpleasant results which could follow the use of compressed air for blowing down clothes. The result was a considerable upsurge in its use apparently because the booklet served as a reminder that here was a convenient way of getting rid of dust on clothing. Beware of guidance consisting mainly of illustrations anyway. Even laboratory workers may be too lax to read the accompanying script. Air showers, which blow a controlled stream of air from above on to clothing to remove dust, are exempted from this criticism.

CYLINDER STORAGE

Here are some basic facts about gases in cylinders. The cylinders are normally of solid drawn steel and must be treated with care. They should not be subjected to shocks, falls, or undue heating. Preferably *all* cylinders should be stored upright. Acetylene and propane cylinders must be kept upright, others may be stored horizontally but never more than four cylinders deep. Where sizes differ the largest should be at the bottom, and securely wedged. An objection to horizontal storage is that it is possible for a mistake to be made so that cylinders which can only safely be stored upright are laid down. In addition, when cylinders are rolled around they are likely to sustain damage particularly to the neck. They also tend to be misused for odd purposes such as propping doors open or as rollers. Upright storage carries the lesser risk of falling bottles but this cannot happen if the fixing devices commonly available are used. Other simple rules regarding storage are that oxygen and acetylene or other combustible gases should be stored separately. Full cylinders should be kept apart from empties, and should be so marked that they can be used in rotation. Grease and oil must not be allowed to come into contact with the cylinders either in store or when in use. The store should be of fire resisting construction and particularly good access is advisable so that cylinders can quickly be moved out in case of fire in the laboratories. Flammable materials must be excluded from the store, and weather protection must be from ice and snow, corrosion, and the rays of the sun. Ideally the store should be facing north and clearly marked showing the nature of the hazards. Purpose-made trolleys should be used for conveying cylinders from the store to the place of use.

There are differing views on the desirability of placing cylinders in use inside laboratories, or outside with the gases piped to the point of use. There are several factors to consider:

Inside Storage

This is less expensive, with relative ease of movement to wherever required and constant surveillance during working hours. There is no risk of leakage from possibly lengthy pipelines, and less possibility of vandalism by intruders who find outside storage more accessible.

Outside Storage

Leakage from cylinders is more readily dissipated outside, and laboratories are not cluttered up by cylinders. Risk from hazardous gases is confined to a single area and, although the risk is magnified in relation to increased quantity, protective measures can be concentrated. Housekeeping within laboratories is made easier, and possibility of impact damage to cylinders or injury from falling cylinders is reduced. If a fire occurs inside laboratories the risk of explosion is reduced. Less frequent changes of cylinders where the usage is high are by the use of manifolds, or bulk containers.

There is a halfway house between the two methods that has merit, which is to provide a suitably constructed cupboard at the entrance to each laboratory in which cylinders can be placed, and the gases then be piped to the points of use. Many factors will govern the eventual choice. Outside storage has most to commend it and in large laboratory blocks is recommended. In medium-sized enterprises cupboards and limited piping is a compromise. Where the use is small then cylinders placed inside are acceptable. In every case the level of safety is governed not only by the physical conditions but also by the degree of care and respect with which gases are treated. A small fire in a technical service laboratory originated in a faulty regulator on an oxygen cylinder. When the cylinder was connected to the manifold it was noticed that the regulator was leaking. The workman concerned tightened the regulator and the leak apparently ceased. The valve was then closed until the oxygen was needed. This did not occur until a week later. When the valve was opened a leak was again identified but before action could be taken ignition occurred, probably because of grease on the regulator. There was also a minor explosion and combustible material in the vicinity was involved. Quick action with extinguishers rapidly controlled the situation. The important lessons arising from this incident were that the regulator probably had a leaking diaphragm. When the leak was first noticed the regulator should have been replaced instead of repair being attempted. As dirt and grease were the probable causes of the leak the need to keep equipment of this kind free from contamination is underlined. Regulators should be returned to the manufacturers for servicing at regular (yearly is suggested) intervals.

HANDLING RULES

Simple rules for the safe handling of liquefied or compressed gases in cylinders should be drawn up, issued, and understood by all concerned in their use. Besides rules peculiar to a specific location dealing with storage (whether piped or otherwise), and so on, certain common principles apply. Examples are:

1. Cylinders inside laboratories should be in an upright position. They should be secured by cylinder stands, by chains affixed to a wall, or on a cylinder trolley.
2. All cylinders should be treated as if full.
3. Gas cylinders should not be refilled unless there is agreement to do so with the supplier, and his approval of the refilling facilities. Normally and preferably cylinders should be returned to the supplier for refilling.
4. Cylinders should not be brought into small unventilated laboratories, or small rooms within laboratories.
5. Cylinder valves should be opened slowly. Improvized tools must not be used for this purpose. First crack the valve open to clear any particles of dirt ensuring there is no risk to personnel or chance of fire while doing so. Then make the connection to the valve outlet.
6. Where there is a manifold linking a number of cylinders ensure a pressure reducing regulator is attached to the header. Single cylinders will have the regulator attached to

the valve. There must be two gauges on the regulator, one showing cylinder content, the other the outlet regulated pressure. Proper maintenance and checks of the accuracy of these gauges must be a matter of routine.

7. Gauges and regulators must be used only for gases for which they are intended and marked. It is common to find oxygen gauges in use on nitrogen cylinders, for instance. Besides the general unsatisfactoriness of this practice, sometimes even condoned by suppliers, it manifests a faulty approach to safety and increases the possibility of errors.
8. Damaged and defective cylinders, valves, and so on must be taken out of use immediately and returned to the manufacturer/supplier for repair.
9. Leaks sometimes occur between the regulator and cylinder. Careful procedure is then necessary. If a flammable gas is involved the cylinder should immediately be removed to a safe place, taking care to avoid possible sources of ignition. If the valve seat is leaking a temporary measure is to attach a regulator to it. Warning notices should be posted, the area cleared, and the contents of the cylinder slowly discharged by gently cracking the valve. The cylinder must not simply be left; the discharge operation should be supervised and controlled by a person knowledgeable about the procedure and attendant hazards. It is sometimes possible as an emergency measure to cure a leak between regulator and cylinder valve by tightening the union nut, but the cylinder valve must be tightly shut before this is attempted. In every case the supplier must be notified and the cylinder suitably labelled and returned in accordance with his instructions.
10. Connecting hoses should be of sound construction, reinforced, and of adequate strength in relation to the pressures in use. Leaking hoses must not be repaired or bound with tape. The defective section must be cut out and the good ends joined by a splice. Legislation regarding cylinders, connections etc. varies from country to country and in some places rubber hose for technical or laboratory purposes is acceptable. Even though this may be legislatively correct, protected/armoured hose is always to be preferred. Entirely unsuitable, lightweight, flimsy rubber connecting hose, frequently poorly connected, has been seen in laboratories. Ends should be fixed to nozzles by suitable fasteners, and should be regularly inspected for tightness and general effectiveness. Hose should be as short as possible in relation to the end use, should not be draped across benches or floors, and should be protected from heat, grease, oil, and impact.

MANIFOLDS

Where gases are piped into laboratories from an outside storage compound the distribution pipework should be designed and installed in accordance with sound engineering standards, and regularly inspected and maintained. Where flammable gases are involved a non-return valve or a pressure reducing regulator should be included at every take-off point to prevent back flow. Small portable manifolds will link up to five cylinders, and stationary manifolds an even larger number. Where oxygen cylinders are linked in this way they must be located away from combustible material and separated from other cylinders containing combustible gases by a fireproof enclosure. Preferably they should be placed in a separate area 16–17 m away.

CODE OF PRACTICE

A typical code of practice concerning the use and handling of gas cylinders and compressed gas supplies actually in use at a large laboratory complex[1] is:

1. Persons changing gas cylinders must be physically capable of handling cylinders and must have been instructed in lifting methods.
2. Before a gas cylinder is disconnected the flow control and cylinder valves must be closed and the delivery pressure adjusting screw turned to its minimum setting, i.e. in a fully anti-clockwise position. (These controls are all shown in Fig. 7.1[2].)
3. Disconnect the cylinder from the pressure regulator, replace the cylinder with the full one, and fasten it securely so that it cannot fall. Cylinders not in permanent racks must be securely fastened in a trolley or free-standing rack.
4. Purge a little gas from the cylinder (not in the case of hydrogen which may ignite spontaneously) and reconnect the pressure regulator. Do not purge toxic gases in a confined space in case a harmful concentration is produced. Remove sources of ignition in the case of flammable gases. Open the cylinder valve slowly, ensuring that the cylinder pressure gauge is operating and then open the valve more fully.
5. Slowly open the delivery pressure adjusting screw until the required pressure is shown on the delivery side pressure gauge. Check that there is no leakage by closing the cylinder valve and observing the pressure readings on each pressure gauge. Open the flow control valve.
6. If no indication is shown on the delivery side pressure gauge after the delivery pressure adjusting screw has been opened widely, the regulator is probably faulty and a new one should be obtained from the stores.
7. By law, regulators must be free from oil and grease.
8. Regulators should be registered and tested every twelve months. It is the responsibility of the user of the equipment to ensure that this is done. Regulators should be returned to laboratory stores for testing.
9. Use only permitted pressure regulators on gas cylinders. In general combustible gas pressure regulators have left-hand thread connectors and non-combustible gas pressure regulators have right-hand threads. Regulators suspected to be defective must be returned to laboratory stores for repair and not be repaired or otherwise modified by laboratory staff.
10. Delivery pressure adjusting screws must be backed off, i.e. set to pass no gas, and cylinder valves closed when the gas is not being used.
11. In the case of cylinders connected to permanently piped installations, the connections must be made by armoured hose connection tubes. These are obtained from gas suppliers and have threaded couplings. They are designed to withstand cylinder pressure. In permanently piped installations only the correct gas cylinders should be connected. If it is required to use a different gas then the colour code should be changed on the line and on the valve.
12. Where gas cylinders are connected to glass apparatus in the laboratory via ordinary pressure tubing, a lute or pressure release device should be fitted to prevent dangerous pressure rise in the system.

13. *Colour coding of gas cylinders*

    Gas cylinder colour codes will help to identify the cylinder contents, but the codes should be interpreted with care. Cylinders of British and United States origin have different codes so that a careful check on the printed name of the gas in a cylinder should always be made. On British cylinders yellow markings indicate toxicity and red markings flammability. The respective colour codes are as follows:

Gas	British Colour Code		US Colour Code
	Cylinder body	Markings	Cylinder body and markings
acetylene	maroon		black/red
ammonia	black	yellow and red	black
boron trifluoride*	aluminium	yellow	aluminium/black
1.3-butadiene*	aluminium	red	pink/yellow/brown
carbon dioxide	black		brown
carbon monoxide	red	yellow	white/red
chlorine	yellow		red
dimethylamine*	aluminium	yellow and red	white/blue
ethylamine*	aluminium	yellow and red	white/green
ethyl chloride	grey	red	red/grey
ethylene	maroon	red	white
ethylene oxide	maroon	yellow and red	aluminium/red
helium	brown		yellow/black
hydrogen	red		yellow
hydrogen bromide*	aluminium	yellow	blue/yellow
hydrogen chloride*	aluminium	yellow	red/brown
hydrogen cyanide	blue	yellow	
hydrogen fluoride*	aluminium	yellow	aluminium/brown
hydrogen sulphide*	aluminium	yellow and red	blue/grey
methyl bromide	blue	black	brown/white
methyl chloride	green	red	red/black
methyl mercaptan*	aluminium	yellow and red	brown/green
monomethylamine*	aluminium	yellow and red	white/brown
nitric oxide*	aluminium	yellow	brown/yellow/pink
nitrogen	grey	black	pink
nitric dioxide*	aluminium	yellow	brown/yellow
nitrosyl chloride*	aluminium	yellow	blue/grey
oxygen	black		blue
phosgene	black	yellow and blue	black/blue
sulphur dioxide	green	yellow	green
trimethylamine*	aluminium	yellow and red	white/yellow

*On these cylinders the British colour code is non-specific and denotes properties only: yellow, toxicity; red, flammability; therefore check the stencilled name.

14. Colour codes are also used for marking valves. These too differ in various countries so a careful check is necessary here also. A typical code but not necessarily international is:

Demineralized water — grass green
Cold water — azure blue
Hot water — French blue (also marked "hot")
Towns gas — canary yellow
Steam — crimson

Compressed air	— white
Vacuum	— green
Nitrogen	— black
Carbon dioxide	— brown
Oxygen	— grey

15. When using cylinders of toxic gases they must be in a well-ventilated area, e.g. a fume cupboard. If the gas is particularly hazardous then appropriate respirator or compressed air breathing apparatus must be kept nearby and ready for immediate use.
16. The flashpoint of a flammable gas under pressure is always lower than ambient or room temperature. When using cylinders of combustible gases, the means of ignition must not be in the vicinity of the work.
17. Special care should be taken when using oxygen to ensure that the operator's clothing is not saturated by the oxygen. If the latter situation occurs, ignition is greatly facilitated and the clothing will burn fiercely.
18. When using chlorinated gases such as Arcton smoking is forbidden, as well as in other defined hazard areas.
19. Cylinders containing liquefied gas, e.g. chlorine, must be used in an upright position and, as in the case of all compressed gas cylinders, in a cool situation. This also applies to acetylene which is present in the cylinder as a solution in acetone. It is unsafe to withdraw acetylene from a cylinder at a rate exceeding one-fifth of its contents per hour; if this rate is exceeded acetone will be mixed with the acetylene. If a greater

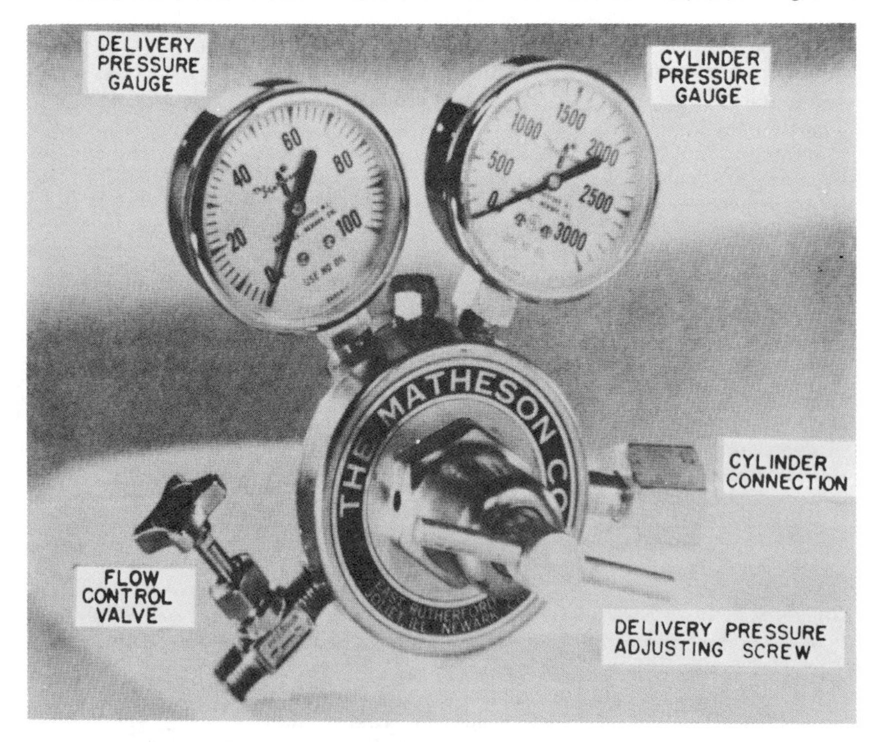

Fig. 7.1. A typical gas cylinder connection showing valves and gauges.

delivery is required then a number of cylinders connected by a manifold should be used. Acetylene regulating valves must not be used on any other gas cylinders. Copper, or alloys containing more than 70 per cent copper, must not be used with acetylene. Cylinders containing liquefied gas are identified by a distinctive stripe painted on the side of the cylinder. Cylinders containing Arcton may be fitted with a regulator which permits gaseous or liquid Arcton to be dispensed.

20. When using combustible gases, flashback arrestors must be fitted between the cylinder and the equipment, if the latter is not already protected in this way.
21. When working with gases piped in from cylinders outside the building so that the line pressure is equal to that in the cylinder, automatic pressure regulators must be fitted to each take-off point. Note that the compressed air supply in (these) central laboratories is at a pressure of 100 psi and that the valves fitted to the standard laboratory bench are of the diaphragm type. These diaphragms stick occasionally and if no air is obtained after the valve has been opened *slightly*, the valve should be examined by a maintenance craftsman. In any event it is necessary to control the air pressure within narrow limits a pressure regulator must be attached to the compressed air point.
22. Treat gas cylinders with respect. A 5.7 m^3 cylinder charged to a pressure of 172 bar (2500 psi) contains the energy equivalent to the contents of a small high explosive shell.
23. This code of practice gives a simplified outline of the precautions to be taken when handling compressed gases in cylinders. More detail is contained in literature (listed at the end of this chapter).

It is re-emphasized that there are differences in the colour code used in various countries so those used in the code of practice above do not apply worldwide. International Organization for Standardization (ISO) R448 used in conjunction with International Standard Book Number (ISBN) 0580 076261 contains guidance on widely used standards.

The fire risk from flammable gas cylinders is high. The use of different threads on cylinder valves, as outlined elsewhere, reduces the degree of hazard but does not diminish the need for care. There are many examples of fatal accidents resulting from failure to follow safe procedures, particularly when oxygen is involved. Three men working on a saw-tooth roof were using oxyacetylene cutting equipment. They went for a meal break leaving the oxygen valve open. When they returned the area in which they were working had become oxygen enriched so that their clothes became saturated with oxygen. Later one of them lit a cigarette and immediately the clothing of all three was engulfed in flames; two died.

Many fires occur because of leaking flammable gas cylinders or hoses. Janitors or watchmen checking laboratories after work has ceased should always check gas cylinders to ensure that valves are correctly closed. Sometimes poor quality or the wrong kind of hose is used, or the hose is damaged. Cylinders fitted with a pressure relieving device present a substantially reduced risk if a fire occurs. Unfortunately many are not fitted in this way so the internal pressure can rise until they burst. Another hazard is accidental damage to the neck or valve which can turn a cylinder into a dangerous projectile.

REFERENCES

1. Tioxide International's Central Laboratories, Stockton-on-Tees, England.
2. Example of common automatic pressure regulator:

FURTHER READING

"Safety in the use of Compressed Gas Cylinders" (1960). BOC International Ltd., London.

"Safety in the use of Gas Cylinders" (1965). Whalley, E. W. F. UK Atomic Energy Authority, London.

"Safe Under Pressure" (1979) (as above).

Code of Regulations for the use of Acetylene. UK Health and Safety Executive, Explosives Inspectorate, HMSO, London.

8

Protective Clothing and Devices

PHILOSOPHY OF PERSONAL PROTECTION

In a manufacturing industry protective clothing and devices are the most costly element in the safety budget. Generally speaking, safety officers and purchasing officers are most concerned with their purchase. Until fairly recently it was commonly the exclusive function of the purchasing officer mainly because first cost was the most influential factor on what was to be bought. Enlightenment has descended in more recent years and the involvement of the safety specialist, the departmental manager, and the supervisor has become more pronounced. This change has been accompanied by a realization that first cost is a most unreliable guide to true cost. Unfortunately, there has not been a corresponding upsurge in expertise in other aspects. Those involved in decision-making in this area of expenditure do not seem to have acquired the degree of competence necessary for wise decisions to be made. The prime consideration is that protection which is adequate in relation to the risk must be selected, followed by an assessment of cost. Personal protection and its selection gives rise to more uninformed comment than most other aspects of safety. In laboratories the proportion of the safety budget expended on it is lower than in the manufacturing industry. The need to get the selection right is still paramount, as also is the desirability of close examination of the economics of the choice made. That is why the subject is dealt with at some length.

Manufacturers and suppliers are not always of spectacular help. Salesmen sometimes try to sell quantities of garments without any in-depth knowledge of weaves, wear life, resistance to specific chemicals, flammability, and so on. Most have undertaken some kind of training, and have technical literature to help, but their all-round knowledge of, for example, how certain materials or garments will perform in stated laboratory work situations and conditions, or of the merits of given weights of cloth in relation to the proposed end use, is normally sketchy. These comments are set down not to deride manufacturers or their representatives but to make the point that purchasers of personal protection should not rely solely on information given to them by those who have a vested interest in the outcome of discussions with a prospective buyer. They should equip themselves with a basic knowledge of what sort of ball game they are involved in. The purpose of what follows is to provide that basis. In many respects it is unspecific: the goal is to make purchasing officers and laboratory managers aware of general principles. Where there is a need for more detailed information, the books mentioned in the bibliography will help.

A frequently expressed assertion relating to personal protection is that it is used only as a last resort. Everything possible must be done to make work safe. The system of work employed, enclosure, substitution, training—all must be examined and brought to a high level of efficiency, and personal protection only introduced when all other methods of avoiding harmful contacts have failed. This statement is only partly true. All these things must be done as a normal part of a laboratory safety programme. The possibility of the unpredictable occurring remains a residual risk that cannot be ignored. Additionally in many countries there is a whole range of legislation requiring specific items of personal protection, and these mandatory requirements must be complied with. Canada has laws regarding clothing to be worn near machinery; Czechoslovakia calls for protective footwear and special clothing for work with noxious materials; Mexico empowers safety committees to decide where and when gloves are necessary; South Africa demands protective clothing where poisonous, corrosive, or other injurious substances are handled. In every country laws concerning respiratory protection exist. This small random selection which consists only of extracts from comprehensive legislation in each country illustrates the universality of mandatory requirements. The need for compliance is paramount.

ASSESSMENT OF NATURAL AND SYNTHETIC FIBRES AND THEIR SUITABILITY FOR LABORATORY USE

The first step in selection is to look at the end use of garments (devices are dealt with later). These questions should be asked:

a) Will dust, corrosives or other special risks be present?
b) Is there a risk of clothing coming in contact with naked flame?
c) In what temperature range will the users be wearing the clothing, and will it be worn near moving equipment?
d) Are flammable liquids or vapours present, or likely to be present?
e) What sort of wear life is looked for?
f) How important is appearance and how often will cleaning be necessary?

In examining these questions in some detail and supplying some of the answers, it will be seen that there is a good deal of interplay between them. This is why the comments which follow are not necessarily directly linked to the questions.

Natural Fibres

The possible presence of dust and corrosives is lumped together because protection factors to look for are similar. The natural fibres, cotton and wool, are good in such conditions, subject to qualifications.

Cotton

Cotton is disintegrated by hot dilute or cold concentrated mineral acids, but is stable in weak, cold acids. It is unaffected by organic solvents and has fair resistance to many chemicals. Not exactly a resounding catalogue of resistance or protection, it might be said, but cotton shares with wool a plus factor not possessed by synthetic fibres—absorbency. So a wearer splashed by a corrosive while wearing a cotton garment has a fair chance of escaping harm because the fibre will, by absorption, lessen the possibility of penetration. When the garment is subsequently laundered, the chance of holes appearing where the fibre has degraded because of corrosive attack is a real one, but the wearer has probably avoided injury or damage to undergarments unless the contact has been very severe.

Other plus factors are good appearance and wear comfort, a high level of wearer acceptability, freedom from static development, good protection against dust penetration, and ease of laundering. On the other hand it will burn (Proban or similar treatment makes it fire-resistant) and does not provide much protection in high or low temperatures.

Wool

Wool as a protective clothing fibre has improved enormously in recent years, due to a considerable extent to the work of the £1¼ million wool research station at Ilkley in the UK opened a few years ago by the International Wool Secretariat. The ability to introduce more bulk, in the form of crimping, into crossbred wools was a development originally initiated for carpet and knitting yarns, but has also found application in the personal protection fields. The introduction of metallic salts of titanium and zirconium has enabled a highly flame resistant variation to be marketed.

Wool is attacked by hot sulphuric acid and affected by alkalis, but gives a useful degree of protection by deflection and/or absorption, plus some natural resistance. When suitably treated, it gives good protection against radiant heat and has fairly low flammability in its normal untreated state. When treated, it has excellent flame resistance. Wool has also a pleasant and acceptable feel, or "handle" as it is known in the industry, has a good wear life, and can be treated to repel molten metal splashes. Static is not a problem. It mixes well with many other fibres, but is difficult to wash, soils tend to cling to the surface, and it shrinks or felts unless cleaned with care. It is expensive these days, but the high regard in which it is held is shown by the readiness of purchasers to dip into their pockets to buy it. If only it were not so hot to wear!

Synthetic Fibres

There is a fairly wide choice of synthetic fibres. Each possesses advantages and drawbacks, as do natural fibres. Thus the point is established that the perfect fibre has not yet emerged.

As a preamble to discussing synthetic fibres, it should be said that some confusion arises because what were originally brand names have now become generic terms used to describe a group of fibres. "Nylon" is a good example. This name was thought up by Du Pont, who defined it as a term for any long chain synthetic polyamide which has recurring amide groups as an integral part of the main polymer chain. Wallace R. Carothers, a brilliant organic chemist, headed a Du Pont team which developed polyamides from initial research in 1928, to the first pair of nylon stockings in 1937, to commercial production which began in Seaford, Delaware, in 1939. Since that time various versions of nylon have been produced in many countries, but generally speaking (there are exceptions) the name "Nylon" has stuck. It is really a polyamide fibre.

Fibres in this group have common characteristics but special considerations apply to some. The aromatic polyamide Nomex is an example, of which more is said later. Nylon is generally resistant to hydrocarbons, ethers, alkalis, detergents, dry-cleaning solvents, and some acids; but sulphuric, hydrochloric, and oxalic acids cause strength losses. Temperatures up to about 150°–160°C will not affect it but above this the fibre loses strength and toughness. It melts in contact with flame and drops away in beads.

Where burns are sustained when nylon garments have been worn, the melted fibre has adhered to the victim's skin and caused treatment problems. Wear comfort is good and so is wearer acceptability. It is an extremely tough fibre having high abrasion resistance, and long wear life in the right conditions is therefore indicated. No difficulty is experienced in dry-cleaning, it launders well if reasonable care is used and its good appearance is retained. It accumulates static and stretches easily under low loads. A fair summary is that, chosen for use in personal protection in the right circumstances, it is quite first class.

Nomex

The aromatic nylon already mentioned, has para-substituted phenylene groups. It has high alkaline resistance coupled with excellent resistance to a wide range of other chemicals not normally associated with nylon. Exceptional resistance to high temperatures is an attractive feature. It will not melt or drip and retains much of its strength up to about 380°C. The soft, pleasant handle and good appearance is retained after continuous launderings. The high abrasion and tear resistance and the lack of effect by heat are factors which put it in the long-life class.

Although increased usage and volume production has gradually made the price more comparable with that of other fibres, it is still dear. But "expensive" is a relative term. Expensive in relation to what? To its injury-saving potential, its long life, and the feeling of added security it gives to wearers? In many high-risk locations in laboratories its cost is justifiable. Motor racing drivers, astronauts, pilots—these and many others have cause to be grateful for its injury and life-saving properties.

Some of the acrylic or modacrylic fibres are sometimes referred to as "the poor man's Nomex". This in no way denigrates them. While not possessing the remarkable protective qualities of the aromatic nylon, they are nevertheless first class for protective clothing in their own right. They have the added advantage of relative cheapness.

Acrylics and Modacrylics

First the difference between acrylics and modacrylics should be explained. Both are polyacrylonitriles, that is to say they are polymers of acrylonitrile which were first developed for spinning into fibres in 1942. It is a little surprising that the potential of acrylonitrile, first produced in Germany in 1893, was not realized for many years. It was not really exploited commercially until the 1930s when it became a constituent of synthetic rubber. In 1945 Du Pont began mass production of acrylic fibres. Acrilan, the best known acrylic, was first produced in 1952. Put simply, a polyacrylonitrile containing less than 85 per cent by weight of acrylonitrile is a modacrylic: at and above this are the acrylics.

Acrilan is unaffected by dilute solutions of strong mineral acids but is attacked by prolonged exposure to concentrated solutions. Strong alkalis will attack it, but it has good resistance to dilute solutions and to common solvents. Its mechanical strength is not impaired at high temperatures and it has no true melting point but will stick to metal at around 220°C. Although it will burn, it is not dangerously flammable. Its "handle" is very good, which accounts for its popularity in a wide range of garments. The wear life, while less good than nylon, is acceptable. It blends well with wool and rayon, can be silicone-treated for water resistance, and accumulates static though to a lesser extent than some synthetics.

There are two modacrylics particularly liked—dynel and teklan. The first is spun from 40 per cent acrylonitrile and 60 per cent copolymer of vinyl chloride. Introduced by Union Carbide Ltd in 1951, it has excellent resistance to acids and alkalis of all concentrations, has very good handle, good abrasion resistance, develops static and originally presented laundering and shrinkage problems. These problems have been overcome and in practical terms it is excellent.

Teklan

Teklan contains about 55 to 60 per cent acrylonitrile and has good resistance to a wide range of chemicals. Warm acetone is one of the few exceptions. It has the additional virtue of very good flame resistance. Almost impossible to ignite, it is self extinguishing should heavy direct flame come into contact with it temporarily. The handle is soft and warm, wearer acceptance is good, it is easily dry cleaned, has good wear life, and mixes well with wool.

Polyesters (Terylene)

Another fibre which arouses enthusiasm is terylene, a polyester. Members of this group are defined as those in which the fibre-forming substance is any long chain, synthetic polymer composed of at least 85 per cent of an ester of a dihydric alcohol and terephthalic acid. A British development, Dickson and Whinfield of the Calico Printers Association, discovered the first polyester fibre in 1941; and development was carried on by ICI Ltd. It has good resistance to most mineral and organic acids and some resistance to alkalis. While resistance to most solvents is good, phenols will swell and dissolve it. It softens at about 260°C and sticks at a slightly lower temperature. Good strength is however retained in prolonged exposure to fairly high temperatures. Although of low flammability it has melting and beading characteristics. Although the handle is "crisper" than nylon due to its lower bending attribute, it is generally acceptable. Only nylon has greater abrasion resistance than terylene. It blends well with cotton and wool, needs minimum care, dry cleans well and has high dry and wet strengths. The development of static can be dealt with by treatment. It has a tendency to "pill", and grease stains are not easily removed, but altogether an exceptional fibre.

Polypropylenes, Polyurethanes, and Rayons

Other fibres to be mentioned fairly briefly not because of demerits but because they have never caught on extensively in the protection field are polypropylenes, polyurethanes, the rayons, and fibre-glass. Propylex, courlene and ulstron are examples of polypropylene. Montecatini of Italy first commercially produced it in 1957. The fibre has excellent resistance to acids and alkalis and in fact to a wide range of chemicals including organic solvents. Its low softening and melting point (150–170°C) is a disadvantage but the main resistance to it on the part of wearers is based on its slightly waxy feeling and some discomfort in contact with exposed skin. It is a strong tough fibre with good abrasion resistance and does not readily accumulate static.

Polyurethane in fibre form is not particularly well known. It has some resistance to cold dilute acids, good resistance to alkalis, solvents and many other chemicals. Its sticking and melting points, though higher than polypropylene, are rather low. It has high durability and exceptional elasticity and has been used in combination with other fibres in the search for a vapour permeable/water impermeable material.

Rayon is a fibre composed of regenerated cellulose and was the first synthetic to be produced on a large scale. The development work on it began in 1892 and Courtaulds developed it commercially in 1904. It forms a high proportion of the total world output of synthetic fibres. Similar in many ways to cotton (also a cellulose fibre) it loses strength at 150°C and decomposes at around 195°C. It has considerable eye appeal and is produced in different forms each having special end uses. High tenacity viscose and rayon treated to give high flame resistance ought to have commanded more attention as a protective garment fibre than appears to be the case. Its popularity will increase as supply factors concerning basic raw materials in other fibres cause difficulties.

Fibre-glass

Fibre-glass is an ancient synthetic fibre. In 1665 Robert Hooks, an English physicist, predicted that glass would be a source of fibre. Louis Schwabe, a Manchester silk weaver, demonstrated in 1842 a machine which produced glass filaments. Economic production began in Owens-Corning Fiberglass, USA, in the 1930s. Its use in protective clothing is largely as the centre or core of a corespun fibre such as used in Heatshield, produced by Courtaulds. This consists of a fibre-glass core which very simply is surrounded by teklan thus ending up as a highly heat resistant and nonflammable material which when woven has other desirable features.

Having decided on the fibre best suited to a particular end use, or uses, the economics should then be considered. First cost is a poor guide to the actual cost of garments. The only true guide is how much a garment costs per month or per other selected period. A cotton laboratory coat may now cost around £10, a similar garment in terylene £18. The cotton garment may last 12 months, the terylene two years. The cost per month is therefore 83p for cotton, 75p for terylene. Pricing only is discussed at the moment; Suitability is another matter. The question of cloth weight is also important —buyers should make sure they are comparing like with like. They may be offered (say) terylene laboratory coats at around £18 from one source and £20 from another. The price comparison is only valid if the fabrics are of the same weight. One would normally expect to pay more for a heavier weight.

CLOTH CONSTRUCTION

The next consideration is cloth construction. The technicalities of weaving and processing were once considered to be of small concern to the protective clothing buyer. Now the importance of correct protection and the wider range of potential hazards have emphasized the need for at least a basic knowledge of these factors. Woven fabrics are produced in lengths commonly known as pieces. Usually the heavier the fabric, the shorter the piece. Width varies, but that most often found in protective clothing use is 90 cm. Weaving is the process which converts fibres into fabrics, the weft threads running across the fabric from left to right, the warp threads running lengthwise from top to bottom. The simplest example of weaving, though in another field, is basket making. The canes can be seen crossing under and over each other, locking together to form a set pattern. The following are examples of the type of weaves usually found in protective clothing materials:

Plain is a method of interlacing warp and weft so that each weft yarn passes alternately under and over a warp yarn—the simplest weave of all often referred to as basket weave or 1 × 1. Sometimes the finished appearance is changed by using a heavier yarn for the weft or warp, or laying two yarns together one way.

Twill. A float is a warp or weft yarn which extends unbound over the weft or warp. This weave is characterized by diagonal lines produced by staggered floats and can simply be described as over two and under one. A heavier, denser cloth is produced than by the plain weave. Normally known as a 2 × 1 construction.

Drill is a variation of the twill weave. The construction is over three and under one, or 3 × 1.

Felt is not really a woven material. Webs of fibre are built up into sheets of required thickness, then "felted" together into a fabric. Often referred to as "moleskin".

CONDITIONS OF PROTECTION

Non-woven fabrics are produced by selecting a mixture of different fibres, the intended end use governing the selection. A fibre with a low melting-point is evenly distributed on to a web of another of the required width; heat and pressure are applied and the fibres bond together as the low melting-point constituent partially melts. Knitted fabrics are produced by creating rows of stitches, each of which hangs on the row behind and the row in front.

Flammables and Static

Where work with flammable or explosive substances is involved synthetic fibres are not normally suitable because of the development of static. They can be treated to suppress this (with Permalose, for example, for Terylene) but it is wise to avoid 100 per cent synthetic fibre materials. In the late 1960s epitropic fibres were developed to get rid of the static problem. The incorporation of 5 per cent epitropic fibre into a material otherwise made up from a polyamide, polyester, or similar eliminated static development. It has never become popular. Mixtures are acceptable and have many desirable features always provided that the mixture is right. Cotton/terylene is an example of how one can obtain almost the best of two worlds. The cotton gives material a degree of absorption, and is considered by many to add to wear comfort; it prevents the development of static if not less than 36 per cent cotton is incorporated. The synthetic fibre constituent adds life and economy to the material. In high risk areas where highly flammable liquids or potentially explosive substances are worked with, test reports on clothing materials used should be insisted on to ensure that they are safe in relation to the conditions in which they are to be worn. One-time or disposal garments have definite applications, besides those in which the destruction of personal protection is desirable if contamination occurs. The high cost of laundering makes them economically attractive in conditions in which they can be worn for a reasonable time before becoming too soiled for further use. There are other factors which add to laundering costs, e.g. collection, despatch, and re-issue of garments. Where disposables can be worn for three weeks or more a feasibility study is justifiable. Normally made up from paper reinforced with cotton or other thread, the material is treated to make it oil and water repellent.

Style

Not enough thought is given to the style of protective clothing for laboratory workers. Traditionally the white coat is standard, but it is arguable whether it is right: A two-piece suit, properly designed, would give much greater protection than a coat. Tradition dies hard, however, and securing acceptance of a change to better protection is not easy. It is frequently the practice to compensate laboratory workers whose personal clothing is damaged in the course of their work. Normally this is paid only if correct personal protection was being worn, and sometimes only if correct procedures were being observed also. The possibility of these claims arising is lessened if a protective jacket and trousers are worn, thus offsetting to some extent the extra cost and adding strength to the main advantage of extra protection and injury avoidance.

Where traditional laboratory coats are worn they need careful design. Fly fronts which completely cover buttons, are essential. Velcro type fastenings have value in selected end uses, but they may be unsuitable where, for instance, strong acids are handled. Cuffs should have fasteners so that the sleeves do not ride up. The sleeve vents should overlap but be long enough to enable the cuffs to be rolled back when hands are washed. There should be no belts or other hanging appendages. Pockets should be positioned and of a size which relates to the work done. Sometimes breast pockets are too small to take notebooks which are frequently needed, and standard equipment. In style too the ultimate in safety is not attainable. Laboratory coats are frequently not completely fastened in front or are even completely open, thus exposing to risk the most vulnerable area. Management by crude disciplinary action often breeds resentment. Sympathetically applied discipline coupled with personal example will partially overcome this practice to which some workers are prone. Functional but attractively designed coats which are well fitting are much more likely to be properly fastened.

It is frequently necessary for workers in industrial laboratories to go into production units, or to work in pilot or technical plants. In this case protection appropriate to the conditions must be worn. Boiler suits or coveralls of a material suitable for the risks are preferred. Laboratory people in a situation in which they are required to work in harness with tradespeople, and others in production or similar areas, sometimes adopt a casual attitude to personal protection and safety requirements generally. This is not always generated by a contempt for the rules. Often it is due to a total immersion in the work in progress so that important subsidiary factors (to them) are overlooked. The reverse should be the case. As often unfamiliar figures to other members of the workforce the example set by them, good or bad, has considerable impact.

The cost of laundering has already been mentioned. Normally garments are collected weekly (sometimes fortnightly) and cleaned by an outside agency. Where the volume is large enough garments from a single source should be laundered in separate batches and not mixed with others from another source. The possibility of difficult soils, non-fast dyes and so on being passed on is then reduced. It may in any case be necessary to separate garments which are contaminated by toxic or other harmful substances. Launderers should be made aware of this kind of contamination. The British Association of Launderers and Cleaners Ltd is a United Kingdom source of information for launderers having complex problems of this nature and it is a helpful precaution to ensure that use is made of it.

It is less common for laundering to be carried out within a laboratory complex. If circumstances are right savings can be made by doing so. Factors to consider are:

a) the availability of a suitable area with plumbing and power source(s);
b) any surplus time among janitors, cleaners etc., bearing in mind that some time will become available anyway because packing, despatch, unpacking, and so on when an outside laundry is used will no longer be necessary;
c) the availability of the necessary capital outlay.

In the smaller location a domestic automatic washing machine, a spin-dryer, and a small ironer will cope with a large throughput of garments. More sophisticated equipment with a larger capacity, together with advice on operation, is available through laundry trade suppliers. Experience has shown that expertise in operating this equipment is quickly acquired. Where domestic machines are installed it will be necessary to specify safety devices, such as interlocks that control agitators when lids are raised, to comply with statutory requirements in many countries. People may be exposed to the hazards of moving machinery in the home, but when the same machines are put to industrial use they must be protected.

EYE PROTECTION

There are basic rules regarding protection of the eyes. They are:

1. The need to protect the eyes arises because a defence system is needed against a hazard which cannot be eliminated, or there is a statutory

requirement to do so. Has everything possible been done to control, or eliminate, the hazard?
2. Comfort and good appearance compatible with proper protection must be considered.
3. The protection provided must be effective in relation to the risk.
4. There must be a comprehensive training programme so that the need for proper protection is understood, as is the importance of correct wear, and care in use.
5. Back-up arrangements are needed so that if an eye injury occurs prompt and effective treatment is available.

Item 1 is self-explanatory. Where hazard control methods have been introduced often there still remains a subsidiary hazard. Where corrosive substances are handled, or there is a risk of explosion/implosion accompanied by an impact possibility, even though conditions appear safe, eye protection is still advisable. Item 2 incorporates careful selection, fitting, and use. Once a need for eye protection has been established then 100 per cent acceptance should be sought and pre-consultation should be followed by general agreement as to its desirability. Representatives of laboratory staffs can then be asked to participate in the selection of the kind(s) to be worn. A variety of eye protectors, each suitable with regard to the risk to which wearers will be exposed, should be obtained. Volunteers from each working area are asked to wear the protectors for one month, then complete a simple form on which is recorded their experience. Types most favourably received can then become standard issues. It is not satisfactory to make available only a single type. There should be a limited choice within the range of suitability. As comfort will be a factor considered in the trials, the selection will take this into account. Individual fitting should follow. It is unwise merely to hand eye protection to wearers in the hope they will adjust it properly and somehow make it fit. Suppliers will provide a fitting service. Good appearance cannot always be achieved. It is possible with spectacle type protectors and they can be as attractive as prescription spectacles.

In relating protection to the risk, accident records will help in identifying hazards and the experience of others can also be used. Where ionizing and microwave radiation sources exist they must be properly shielded and personal protection not relied on. There may be occasional exceptions to this when leaded lens are needed. Where lasers are used an essential element of control is the wearing of effective laser eye protection. Different filters are needed for various laser systems and all should be clearly marked and identified by different colour frames. The protection will have maximum attenuation at a specific laser wavelength, and should be obtained from eyewear manufacturers strictly in accordance with the end use. One-piece goggles, classified in British Standard 2092 as "chemical eye protectors" and having a similar classification in some other national/international standards, are the recommended wear for general laboratory use, because the main injury potentials are chemical splash or glass breakage. Goggles are subjected to a robustness test, but where there is a fairly substantial impact risk a mechanical strength test should be specified. Full face protection may be needed in high risk situations, particularly where spectacles (with side frames) are worn. There is a perpetual argument as to the desirability of the spectacle type eye protection fairly common in laboratories. Those who favour it argue that it is more acceptable to most workers and is therefore more likely to be worn continuously than the less comfortable chemical goggles. There is a decision to be made here by management after working conditions have been carefully assessed. Bear in mind that worldwide 5 per cent of all indus-injuries are to the eyes, and that 50 per cent of all eye injuries treated in hospital occur at work.

Training with regard to protecting the eyes should be aimed at convincing users of the need; demonstrating types available; explaining statutory requirements; and showing cleaning, maintenance, and general care procedures. Back-up arrangements consist of small wall-fixed cabinets containing cleaning and anti-mist fluids, and cleaning tissues. Eyewash bottles and eye irrigation systems should be strategically placed and clearly identified. First aid rooms, medical departments, or whatever form of secondary treatment is available after on-the-spot first aid has been rendered, should be equipped to deal with, and have personnel knowledgeable about, eye injuries. Liaison with the nearest eye hospital will embrace creating an awareness of specific risks in the laboratory, seeking and taking advice from the specialists there, and ensuring that rapid transfer to hospital is always possible.

HAND PROTECTION

Hand protection for laboratory workers is normally confined to that which will prevent corrosives or irritants coming into contact with the skin. The few exceptions are dealt with first. The minor need for material handling purposes can be met by leather gloves, usually a cow or horse-hide split-tanned by a chrome or chrome-vegetable process. This tanning process has significance because when the chrome content is raised the gloves, which are the end-product, will have considerable heat resistance and can be used in conditions where asbestos gloves were formerly the only choice. The objections to using asbestos in any form makes these "thermo-leather" gloves particularly acceptable. Cotton gloves have limited uses but should not be used where protection from, for example, broken glass is needed.

Protection against harmful substances is provided by PVC, neoprene, nitrile, or rubber gloves. Polyvinyl chloride is stable in chemicals and oils, will not break down in temperatures up to 120°C, is waterproof, has a high resistance to abrasives, and is easily cleaned. It can be moulded into seamless gloves, or supported by fabric which may range from stockingette to a lightweight synthetic material. Its disadvantages include embrittlement, which can occur at low temperatures. It is subject to penetration by sharp or rough objects; and does not have a high impact resistance. Where exceptionally hazardous substances are to be handled, the gloves can be subjected to a pressure test to ensure impermeability. Made up in a rough finish or a rib-grip pattern, these gloves will stand up to quite heavy material handling tasks. Less good than leather in this application, they are nevertheless preferred if materials to be handled are irritant or corrosive.

Neoprene Gloves

Neoprene rubber is polymerized chloroprene. It is produced by reacting two molecules of acetylene to form vinyl acetate which is then reacted with hydrogen chloride. It has a high melting range, is chemically inert and is resistant to friction. In low temperatures it performs much better than PVC. A factor noted with wearers of neoprene gloves in hazardous conditions is that they claim an added sense of extra protection. In many ways neoprene is similar to nitrile rubber.

Nitrile Gloves

Co-polymers of butadiene with varying amounts of acrylonitrile form the nitrile rubbers. They are notable for high resistance to oils, but have been found satisfactory in a range of uses. The sense of touch is mainly affected in all gloves by the thickness of the material, but there are other factors. Wearers of nitrile gloves claim that tactility is greater than with other gloves of similar thickness.

Rubber Gloves

Before substitutes were available, these gloves enjoyed a wide use. They are still used by, for example, surgeons, cleaners, and electricians. A disadvantage is their tendency to cling to the skin and promote sweating, although flock linings alleviate this. When used as protection for people working on electrical units they should comply with the requirements of British Standard 697: specification for rubber gloves for electrical purposes, which contains four classes according to rated voltage, requirements and tests for electrical resistance and mechanical properties, and recommendations concerning maintenance and retesting. There are similar standards in most other countries.

Asbestos Gloves

Mixed with wool as a woven material then made up into gloves, asbestos has a good deal of merit. Unfortunately, the reaction against this mineral has led in some cases to a refusal to wear these gloves. They are highly resistant to heat, and when surfaced with reflective aluminium foil can be worn in extremely high temperatures.

When caustic substances, acids and harmful solvents are poured from large to small containers, gauntlet type gloves and sleeves worn outside the gauntlet are needed. Where particularly hazardous substances such as tetraethyl lead are handled there should be a daily visual inspection of gloves accompanied by a pressure test. All gloves should be routinely inspected and cleaned. It is a common misconception that leather gloves cannot successfully be cleaned. This is not so. An ordinary water/detergent solution used at about 48°C will remove most exterior dirt and also accumulated sweat from the inside. Before drying it is helpful to rub over the exterior with a cloth lightly impregnated with paraffin. Drying should be slow and at a moderate temperature.

When gloves, face shields, and other personal protective equipment becomes contaminated with harmful substances a careful removal procedure should be set up. Before it is removed it should be thoroughly washed with a water stream. Coats and aprons should be removed first, then gloves. The hands should then be washed, and eye protection removed. All the equipment should then be carefully washed again to remove any remaining contaminants, and gloves sanitized by the use of a mild disinfectant.

RESPIRATORY PROTECTION

Careless use and handling of respiratory protection has been noted in laboratories. This kind of personal protection is so important an item that strict procedures relating to its use should be devised and insisted on. General rules are:

1. Make sure the protection is right in relation to the risk.
2. Check that it is in good order, and that cylinders etc. and contents have enough capacity for the work to be done.

3. Ensure that wearers have had adequate and reasonably recent training.
4. After use, a systematic return to a central control and maintenance point for cleaning, checking and re-issue in the case of personal issue.
5. Simple but accurate records of each item, showing periods of use, dates of examinations, repairs and renewals.

In larger laboratories there will be a full- or part-time safety officer who should be familiar with the respiratory hazards present and the type of equipment suitable for each risk. He will oversee the strict application of laid-down procedures and not permit any deviation from them. In smaller laboratories a chemist or technician can undertake this function after suitable training. Respirator types are discussed below.

Canister Respirators

Canister respirators consist of a face-piece connected by a corrugated flexible tube to a canister. Contaminated air is drawn in through a one-way valve, passes through the chemicals in the canister, then via tube to the face-piece in a purified state. The contents of the canister must be related to the risk. General-purpose canisters exist that give protection against a number of contaminants. Although they have value in circumstances where a plurality of gases or vapours may be present, they have a shorter life than those related to a specific class of contaminant. Respirators of this kind can safely be used only when there is a minimum of 16 per cent oxygen present and not more than 2 per cent of a contaminant, in both cases by volume. There are some variations to this rule in relation to the percentage of contaminant, e.g. the limit for phosgene is 0.5 per cent. Charts are available from manufacturers of respirators and canisters, and there is also a number of national standards, to which reference should be made. Factors governing the effective life of the canister are the degree of activity in which the wearer is involved, the type of canister, and the concentration of gas or vapour. A rule of thumb is that after one and a half hours' wear a canister should be renewed. When the useful life is running out anyway small indications of the contaminant present passing through the canister unfiltered will warn the wearer. If the gas or vapour is so highly toxic that the wearer would be affected by this indication a respirator of this type should not be in use anyway—self-contained breathing sets should be the choice.

Canisters are sealed while in store to prevent absorption of contaminants. The seal is broken when the canister is attached to the tube and facepiece, and long storage thereafter may lead to absorption of chemical vapours. Enclosure in a vapour-proof plastic bag will prevent this. A final point about this kind of respirator is that, because it is reasonably light and comfortable to wear, there is a tendency to use it when more sophisticated equipment is needed. This applies particularly to self-rescuers, or pocket respirators, intended to give very short-term protection to allow escape from a contaminated to a clean atmosphere. Canisters also have a limited shelf life in store.

Dust Respirators

Dust respirators have a chemical filter that will give protection against dust. The filtering medium must be related to the particle size of the dust present. The more efficient the filter, the greater the resistance to breathing becomes. This can be overcome by enlarging the area of the filtering medium. Filters which are relatively ineffective because of high porosity and consequent low resistance to breathing should be avoided, although they are popular because of wearer comfort. Filters must be renewed when they become plugged or "blind".

Air-supplied Respirators

Air-supplied respirators enable the wearer to work quite independently of the surrounding atmosphere. Familiar types are hose masks with a long tube or blower in respirable air (of limited use in laboratories), air line masks which connect by means of bayonet or similar type of rapid attachment to a source of clean air under pressure—also of rather limited value in laboratories. Self-contained breathing apparatus gives complete protection to the wearer irrespective of the concentration of contaminant present, allied to complete mobility. Breathing apparatus of this kind may be of the oxygen regenerative type, or have compressed air in a cylinder. The latter kind is simple to wear and maintain, and is generally favoured. It consists of the air cylinder (the size of which is the main factor governing the wear life), a valve, a corrugated tube, and a facepiece fitted with a demand valve. The wearer dons the facepiece, adjusting the head harness to ensure an air-tight fit. The cylinder valve is then opened, and thereafter draws air through the demand valve as required. A gauge accessible to the wearer shows the amount of air in the cylinder. The duration of these sets is a factor governed by legislation in many countries.

It is again emphasized that the price of personal protection should be realistically assessed bearing in mind that first cost has an imprecise relationship with true cost. Simple records will indicate the wear life of each item and a simple calculation will then translate first cost into cost per month. In departments where there is exposure to X-rays or gamma radiation leaded clothing is required or electromagnetic radiation suits for high level radiation. These specialist applications should be dealt with in conjunction with manufacturers of apparatus and personal protection.

Decision-making in the area of protective clothing and devices is just as much the province of the laboratory manager as is the design and staffing of laboratories, the substances used, and programmes of work. The prime consideration is that workers must be properly protected. Then follows decisions as to how this aspect of the safety budget can most wisely be spent. It is a false philosophy that the safety purse should be bottomless. Skimping on expenditure at the expense of safety is inexcusable. So also is the unnecessary pouring down the drain of an enterprise's assets simply because managers cannot take the trouble to combine effectiveness with economy.

NOISE

Occupational deafness has received belated acceptance as an industrial disease. Exposure to noise levels likely to be harmful is less common in laboratories than in many other work locations. Nevertheless as well as the moral duty to protect employees' hearing there are in every country statutory requirements governing maximum permissible levels. In most countries this is expressed as 85 decibels at frequency on the "A" scale (85 dBA) for eight hours exposure. Noise suppression at source is the ideal. Where this cannot be achieved personal protection can be through ear muffs, or plastic or glass wool plugs.

FOOT PROTECTION

There may be a minimal need for foot protection in laboratories. Shoes with steel toecaps complying with the applicable national standard should be chosen. A neglected factor in protective footwear is the soling pattern. Many more injuries are caused by slips and falls than by objects falling on unprotected feet. Footwear with soles having the least tendency to slip should be chosen.

AIR SHOWERS

The removal of dust from clothing can be achieved by air showers. A typical arrangement is a chamber having an entrance and exit door. The person wearing dust contaminated clothing enters the chamber. The action of closing both doors activates a time switch that motivates an overhead air pump which ejects streams of air downwards, blowing dust from the clothing of the person standing below. At the expiration of a predetermined time the airstream is cut off. The showers have value in, for example, pharmaceutical laboratories by removing dust and lint from clothing before an "uncontaminated air" area is entered. The air in the chamber is extracted via trunking, filtered, and then discharged to atmosphere. Air showers have the disadvantage of failing to remove clinging dusts so that the blowing action needs to be supplemented by mechanical action such as brushing with the hands. In any case lint-free clothing should be specified where uncontaminated air is important, and clothing worn in laboratories should be discarded before canteens are entered. Air showers still have a place in certain laboratories but their limitations should be recognized.

FURTHER READING

Cockett, S. R. (1966). "An Introduction to Man-made Fibres", Sir Isaac Pitman & Sons, London.

Cook, J. G. (1968). "Handbook of Textile Fibres–Natural Fibres", Morrow Publishing Co., Watford, UK.

Davies, C. N., ed. (1962). "Design and Use of Respirators", Pergamon Press, Oxford.

Freeman, N. T. (1962). "Protective Clothing and Devices", United Trade Press, London.

Hall, A. J. (1975). "The Standard Handbook of Textiles", Newnes-Butterworth, London.

Labarthe, J. (1975). "Elements of Textiles", Collier-Macmillan, London.

Moncrieff, R. W. (1975). "Man-made Fibres". Newnes-Butterworth, London.

"Protection of the Eyes" (1963). Chemical Industries Association, Alembic House, Albert Embankment, London.

Rousell, D. F. (1979). "Eye Protection", Royal Society for the Prevention of Accidents, The Priory Queensway, Birmingham, UK.

SOURCES OF SUPPLY

Draeger Safety, Sunnyside Road, Chesham, Bucks, UK.
Efficiency Aids Ltd, 300 Norton Road, Stockton on Tees, UK.
Walter Page (Safeways) Ltd, 46 Lower Shelton Road, Bedford, UK.

9

Fire Protection and Prevention

PRINCIPAL CAUSES OF FIRE

It is an apparent contradiction that fire losses rise annually in real terms while at the same time the fire protection of structures, fire separation, automatic detection and extinguishment, and similar matters become more sophisticated and effective. The reasons for this anomaly are varied and a brief review of them is rewarding. Attention to the principal causes of fires is important: know the enemies then take all reasonable measures available to combat them. An analysis of 1210 £15000-plus fires in the UK[1] showed that 45 per cent of fires for which the causes could be determined were started deliberately by adults or children. Next came electrical equipment, followed by smoking materials, then naked lights. There is also a behavioural problem in fire safety as in other aspects of safety and the comments made elsewhere apply in this context also. Undisciplined conduct, pushing lighted fireworks through letterboxes as a "joke", tossing matches into rubbish containers in the rear of laboratories: account must be taken of these and similar possibilities. The investigation showed that more fires broke out in storage than in working areas and 60 per cent of the fires occurred outside normal working hours. An oddity revealed was that damage resulting from fires in modern premises was more expensive than that in older premises.

An important contributory factor is that modern premises (modern was defined as 30 years old or less) are usually designed with ceiling voids housing electrics, air-conditioning and fume extraction trunking, and other systems. These are hidden from view so that a fire can develop undetected until it has secured a fierce hold. In older premises many services have been added after the original construction and can be seen. When a fire prevention/detection programme is being developed account should be taken of these considerations.

The overall control of fire risks is such a highly complex and technical matter that the details are almost infinite. In a book of this kind it is only possible to set down the framework on which an intelligent fire control programme can be built. In small locations what follows may need little padding; in large laboratory blocks expertise within the enterprise, or brought in from outside, should be applied to produce a programme tailored to specific needs. A feature of laboratory fires is that the actual damage to materials and structures may be relatively insignificant when compared with such losses as records, results, and computer discs. Ideally, important material should be duplicated and the copies stored in another building. In practice this is seldom done. The less satisfactory alternative of storage in fireproof cabinets is more common. Fireproof in this context is a relative term. These cabinets will protect the contents in minor fires. After major fires they are often found undamaged but with the contents "cooked" or destroyed by heat. In any case document and computer programme safety can only be secured by this means where a routine is developed, accepted, and carried out inexorably. The fallibility of humans being what it is, this only occurs where determined management exists and rigidly enforces the routine. Recognizing the need to safeguard laboratory records from destruction by fire and as a shield against industrial espionage all standard filing cabinets were removed from a suite of laboratories. At considerable expense they were replaced by lockable four-drawer cabinets which were "fireproof". A few weeks after this change a security officer carried out a special check on the cabinets outside normal working hours. Threequarters were unlocked and nearly one in five had drawers open and one or more drawers left extended. It is easy to say "That would not happen here": experience shows that it can and does happen in a very high proportion of locations. Before exempting your laboratories, have a check one evening or weekend.

DESIGN AND FIRE SAFETY

Laboratory managers are not always involved in the design of new buildings or the rehabilitation of older buildings which they are later to occupy and control. They should be an important component of the design team. It is at this stage that the fire programme has its beginnings. Managers are aware of the kind of work to be carried on in each laboratory, the substances to be used, their flammability and toxicity. They can advise on the extent of fire separation needed, the nature and protection of means of escape in the light of numbers employed, and the kind of built-in fire extinguishment that is not only desirable but safe (e.g. should sprinklers be installed in laboratories in which substances used are reactive with water). The fire protection of structures, the application of general principles leading to the grading of buildings according to the fire resistance of their elements of structure, and similar matters are the subject of governmental and/or local authority control.[2] Designers are aware of these requirements and the need to comply with them. The function of managers is to draw attention to the operational factors such as end uses and numbers employed so that proper account can be taken of them at the design stage.

FIRE PROGRAMME

Every location should have a fire programme embracing the fundamentals of protection, prevention, and control. The following elements are the components of a complete programme:

1. The preservation of human life.
2. Prevention methods to reduce the possibility of fire occurring.
3. Provision of means for the early detection of fire, or raising the alarm and transmitting that alarm to a place where action can be initiated.
4. Adequate first aid fire-fighting equipment to facilitate prompt extinguishment where practicable.
5. Prevention of fire spread.
6. Minimization of fire damage, including damage from water and other extinguishers.

Preservation of Human Life

Preservation of human life is the first priority. Equipment, buildings, and records can be replaced, lives lost cannot be regained. The establishment and routine practising of an approved procedure to be followed in case of fire is the first step. "Approved" means discussed with and finally approved by the fire authority for the district. A fire alarm system tailored to the needs of the laboratories is the starting-point. In a single laboratory or other small locations a sophisticated alarm may not be necessary. A hand-operated bell which can clearly be heard may be adequate. The important consideration is that the means of raising the alarm is clearly and unmistakably heard on every part of the site. Where more advanced means are needed these may be manual systems which entail the operation of a manual fire alarm switch, or automatic systems which detect a fire and operate an alarm without human intervention. In either case an annunciator or control panel is needed. This should be located in a continuously manned position if this exists, otherwise in a place where it can easily and clearly be seen. The panel will show the laboratory or floor on which the alarm has originated. Automatic systems, normally installed in medium/high-risk buildings, should be linked by direct line to the appropriate fire station or fire control room. Alarms of any kind should be tested regularly—once a week at a predetermined time is reasonable—any defects noted, appropriate remedial action taken, and the detail logged. Manual alarms should be activated from a different switch each time, and there should always be a fall-back on to the battery system to take account of power failure. According to the location, opened and unsupervised, or closed and electrically supervised circuits may be used. In the first case, power failure or a circuit fault causes the alarm to operate, in the second case a visual fault signal will show on the control panel but the alarm will not operate. The latter is preferred. Open circuits were installed in laboratories attached to a large hospital. The fire alarm operated on three occasions in one week because of circuit faults. Not only were the laboratories evacuated, which was a nuisance, but evacuation of the hospital began before the fault was identified. In the case of seriously ill patients, tragedies could have occurred. The system was changed to closed-circuit at a cost of around £65000. The importance of getting it right in the first place is thus emphasized.

The alarm having sounded, the procedure to be followed must be set down, clearly understood, and followed without question. The last part of this sentence cannot be too strongly stressed. For example, it is common sense to transmit the alarm to the fire authority immediately where it is not accomplished automatically. This should be part of the procedure. Too often it is not carried out, sometimes with disastrous consequences. There should be no intervention of a decision-making process. When fire occurs, call the fire brigade. This procedure is accepted and approved by the fire authorities throughout the world. Fear of repercussions if the attendance of the fire brigade proves unnecessary is unwarranted. There simply will not be any.

The wording of the fire procedure is important. In some laboratories notices setting this down are so long and laborious that five minutes reading is entailed. The wording should be as short and concise as possible. Try setting down a procedure, then reduce the number of words by half, then try a further reduction until the irreducible minimum is attained. The result may be something like this:

FIRE

1. OPERATE ALARM.
2. IF SAFE TO DO SO, ATTACK WITH EXTINGUISHERS.
3. IF NOT, TURN OFF GAS AND ELECTRICAL APPLIANCES—EVACUATE.

IF YOU HEAR ALARM

1. EVACUATE USING QUICKEST ROUTE TO EXIT.
2. GO TO ASSEMBLY POINT.
3. REMAIN THERE UNTIL INSTRUCTED OTHERWISE.

Clearly the procedure must be linked to the establishment of well-defined assembly points. These should be marked out and located at a safe distance from, but as near as practicable to, the laboratories. Escape routes will be defined by the fire authority. They must be known to all and must be maintained in the condition laid down. The following factors are taken into account.

Stage 1. Distance of travel within laboratories or attached offices.
Stage 2. Distance of travel from laboratories etc. to a stairway or final exit.
Stage 3. Travel within stairways and to final exit.

Useful guidance is that where there is escape in one direction only the distance to the exit from a laboratory from any point should not exceed 12 m. The distance of travel to the final exit or a protected stairway should not exceed 45 m if there is escape in more than one direction, or 18 m if escape is in one direction only. These distances are inclusive of distances within a laboratory. The width of exits should normally not be less than 750 mm and corridors 1 m. This detail is given because it is an important function of laboratory management to see that these distances and dimensions are maintained. Stores or equipment are sometimes stacked in escape corridors so that the width is reduced by half. Exit doors are locked permanently where this is considered necessary for security reasons instead of making use of panic bolts or latches, or some other acceptable means of fastening so that they can be opened readily by persons escaping. Barriers, benches, or equipment are placed inside laboratories so impeding or completely obstructing free access to the exit door. Day-to-day attention to these matters is positive, effective, and infinitely preferable to the lamentations and regret that follow tragedies resulting from unusable or obstructed means of escape.

Because all employees must be aware of the fire procedure a routine should be established to ensure not only that this is attained but also that regular practices provide a constant reminder. New starters on their arrival in the laboratory should be given a verbal introduction to procedures, supplemented by a concise handout to be retained. They should be conducted along escape routes and shown their assembly points. Evacuation drills should be held at six-monthly intervals, each alternate drill incorporating a prearranged attendance of fire appliances. When the drill is completed the appliance crews should be taken on a familiarization tour of laboratories. These drills can be carried out with the minimum interruption to work in progress. The effective minimum number of staff should be detailed to remain at the place of work when the alarm sounds. They will monitor equipment/experiments which need to be left running. When the drill is over and normal work resumes, staff who have not participated will be conducted over escape routes to their assembly points and interrogated as to their knowledge of the over-all procedure. The names of absent staff should be noted, and on their return to work they too should participate in the routine suggested. Only in this way can genuine 100 per cent coverage be assured.

Fire wardens can play an important part in the fire procedure. At least two, normally unpaid, volunteers should be appointed in each area, the size of which should be governed by the degree of risk, quantity of flammables, and similar factors. Wardens need simple training in the use of extinguishers and hose reels, and also in the functions they are to perform. A simple handbook containing concise details of their role should be issued, as should some means of identification such as an armband or distinctive helmet. Briefly, their functions should be to:

a) Familiarize themselves with the position of fire alarms, hose reels, and extinguishers in the section to which they are appointed, the means of escape, and ensure that new people know the fire procedure.

b) Operate the fire alarm if this has not already been done when a fire occurs. Attack it with extinguishers if it is safe to do so, and check that evacuation in their section is complete. If time permits close all doors and windows. Go to assembly point, check personnel, and report to fire brigade "all present" or name any missing person(s) and where last seen. Keep evacuated people together and supervise an orderly reoccupation of laboratories when authorized to do so by the fire brigade. After each evacuation drill or actual alarm wardens should be called together for a short discussion to identify and remedy any problems which may have arisen.

Special problems are posed where a flexitime system is operated, and exacerbated where there are also frequent visitors. It is difficult for fire wardens to know at any one time who is present in their area. There are systems which will overcome this difficulty. One example consists of an indicator board showing the laboratories in plan, and each area for which a fire warden is responsible (see Fig. 9.1). Located in each area is a switch. When the fire alarm is operated the board is activated, the flashing light on the left-hand side showing that the system is live. Fire wardens check their areas to ensure that evacuation is complete and made as fire-safe as possible; they then operate a switch as they leave, which lights up the related section on the indicator board. A quick glance at this shows whether any areas have not been checked so that the fire crew and fire brigade can initiate search and rescue. The system is fail-safe so that if a fault develops the area involved will be recorded as unchecked. In practice this device has been found to work well.

The location of alarm operating points should be adjacent to the exit from laboratories and clearly marked. The establishment of "Fire Points" is recommended. These may consist of a triangular arrangement on a wall beside the exit. At the apex is the alarm-operating switch. Immediately below are the procedure instructions, then hose reel(s) and extinguishers suitable for the risks in the area. Extinguishers should be hung on brackets with the handling part 1 m from floor. The Fire Point should be clearly marked, perhaps by an 8 cm-wide painted triangular band, enclosing the total installation. The location is chosen because if a fire occurs those working in the area, following the first principle of personal safety, immediately go to the exit door. The alarm is operated then the fire viewed. Only if it is safe to do so, extinguishers or hose reels are then used to fight the fire, other-

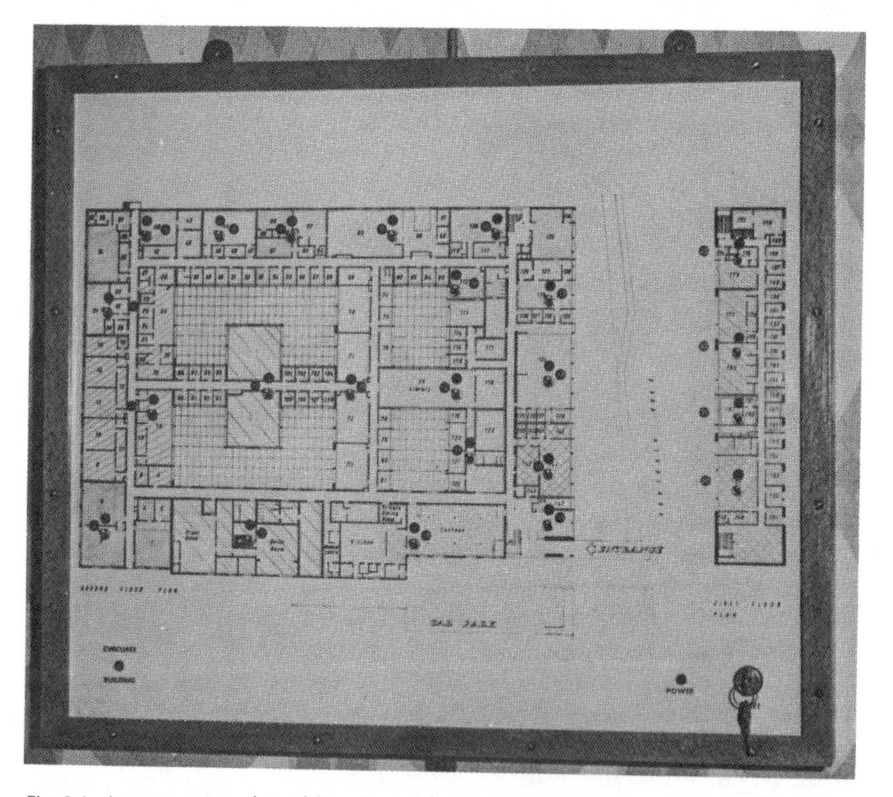

Fig. 9.1. An annunciator board for a large laboratory block. The board is connected to the fire alarm system so that when the alarm sounds, the board is activated. This is indicated by a flashing light in the bottom left-hand corner. As fire wardens check their areas and leave the building they press down the switch for their area which lights up an indicator bulb. Inspection of the board by the fire team leader tells him whether any part of the building has not been checked and he can then arrange a search party of the unchecked areas.

wise the general evacuation is joined. Looking at this situation in another light, if a fire occurs in a temporarily unoccupied laboratory, a passer-by seeing the fire does not have to go into the fire area to raise the alarm and perhaps attempt extinguishment. The means are readily at hand. (See Fig. 1.10.)

Prevention Methods

The main causes of fire have been identified. Prevention should take account of them. Security measures are discussed elsewhere, but their importance in the context of fire prevention must be emphasized. If intruders are kept out of buildings their fire-raising propensities are kept in check to some extent. Many fires begin in outbuildings and so they must be protected. Even when buildings are properly designed regarding fire safety and built-in fire protection exists, there should be periodic self-inspections as part of the fire safety programme.

Things to look for are evidence of smoking in non-smoking areas, excessive quantities and careless handling of flammable liquids, which should be kept either in a purpose-constructed flammable liquids store or in smaller quantities in a fireproof cabinet. In any case only the amount needed for immediate use should be on benches. It is quite usual to see whole drums of highly flammable liquids in laboratories where the usage is only a litre or so each day. Other defects which may be found are bad housekeeping, accumulations of carbonaceous materials, defective electrical equipment, distribution boards and fuseboxes left open, overloaded or badly sited electrical outlets, and ring main systems where highly flammable liquids are present. In these systems an arc is drawn each time a plug is pulled out of a socket. Laboratory bench surfaces may not be protected from overheating, hotplates used to heat flammable solvents, and failure to use fume cupboards when necessary.

These important features of laboratories tend to be little used, or grossly misused. Often they are cluttered up with bottles left by previous users, dust-covered apparatus is left within, and the exhaust ventilation is allowed to become inadequate. The eventual external outlet of fume cupboard extraction ducting should be examined. Defects noted include outlets at right-angles to an external wall on which the prevailing wind blows so that on windy days fume is blown back into the laboratory, and a duct terminating on the landing of an external fire escape. In another case where fume cupboards were used mainly in connection with flammables the outlet ended about two feet from the ground level in a car park: cars were backed right up to the outlet. Other risks include gas apparatus left burning and unattended, gas cylinder valves not properly closed, and standard refrigerators which may present a fire/explosion risk. Bunsen burners are still extensively used in laboratories, and if carelessly used they present a fire hazard. There are many examples of problems arising from them. Forceps flamed by a Bunsen burner but not allowed to cool enough, once caused a flash fire in an alcohol beaker, and in the excitement the technician tipped over the alcohol. In another case a technician rinsed his hands in alcohol and came too close to a Bunsen burner. The alcohol ignited but serious injury was avoided when the fire was quickly smothered. Another technician, inoculating eggs in a sterile room, upset a beaker of alcohol near a Bunsen burner. Flaming alcohol covered her clothes, but fortunately her partner was able to smother the flames. In a hospital laboratory in the UK acetone was decanted into a beaker from a larger container. The telephone rang so the technician put the larger container down unstoppered near a lighted Bunsen, and then moved off to answer the telephone. The vaporizing acetone was fired with explosive force. Distractions of any kind must not be allowed to interfere with safe procedures. Simple rules coupled with sensible operation, will avoid mishaps with these burners. Do not buy cheap equipment. Make sure the burner is designed for the heat value of the fuel gas. Keep the burner tip clean to prevent the containment of concentration by back pressure, and do not use home-made apparatus. Hotplates are notorious for igniting low-boiling point solvents because the plate element temperature is often above the ignition temperature of the solvent and because the arc of the make-and-break thermostatic control is hot enough to ignite the vapour. One technician chose to use a hotplate to hurry the evaporation of ether. The ether caught fire, but fortunately the flask did not break or tip, so the resultant fire was fairly easily extinguished. In yet another case petroleum ether, which an investigator was pouring from a glass container, was ignited by a nearby hotplate which had recently been turned off. The researcher dropped the container and ran from the room, having suffered hand burns. The ignition of the total spill resulted in considerable damage.

Means for Early Detection

Care in selection of fire detection systems will not only help to ensure that the right system is used in relation to risks, but also that operation in circumstances other than a fire is minimized. Ill-chosen systems frequently operate when no fire is present so that in time the "it's just another false alarm" mentality develops and the entire fire programme may fall into disrepute. For example, rate-of-rise detectors in a small room in a laboratory complex were triggered off when painters used calor gas torches to burn off old paint. The rapid rise in temperature which followed activated the detector heads. In a laboratory in the same building the central heating was reduced on

Friday evenings and stepped up again at 7am on the following Mondays. On a particularly cold Monday morning laboratory workers, attempting to step up the temperature, lit two drying ovens and left the doors open. Immediately overhead was a rate-of-rise detector which was activated by the rising thermal currents, and yet another false alarm occurred. A smoker lit his pipe in a conference room and created a cloud of smoke which set off an optical smoke detector above his head. Many of these and similar incidents can be avoided if people act in a sensible and thoughtful way, but often they do not. Below is a brief description of the common detection systems. Laboratory people may or may not be involved in the choice made. Some knowledge of how they work will help to reduce false alarms.

Ionization smoke detectors give early warning of smouldering or open fires. Their advantage is that they heighten the possibility of successfully fighting a fire, because the warning is quite frequently triggered off when fires are in the incipient stages. The possibility of operation when there are no fires is equally enhanced. Optical smoke detectors react to visible products given off by most fires in the early stages. They are a reliable type of detector which nevertheless must be sited with care. Another type of ionization smoke detector reacts to invisible, as well as visible, products of combustion, has the advantage of detection before flames appear, and is available in a range of sensitivities. Infra-red radiation detectors react to the emissions from flickering flames. Their particular applications are in laboratories with high ceilings or where smoke may be given off by the work carried on, in which circumstances smoke detectors are of limited effectiveness. Rate-of-rise detectors operate when the ambient temperature at the detector increases. They may be activated when a pre-fixed temperature is reached, by a rapid rise in temperature, or by both. The sensitivity of some detectors can be adjusted to take account of the conditions in which they are to be used. Some incorporate an unauthorized removal indicator. If this is thought to be an unnecessary refinement, consider that it is not unknown for employees to remove a detector which has become something of a nuisance by operating the alarm when fire is not present. This situation should be dealt with by adjustment of sensitivity or change of type, bearing in mind that many detectors are interchangeable with others. Careful checks and experiments with detectors in co-operation with the fire authority will enable a final setting to be arrived at which will give maximum protection coupled with the unlikelihood of maloperation, provided that people working in the area behave sensibly.

The use of security guards is mentioned elsewhere. They form the oldest type of fire detection. It is indefensible to leave unprotected for up to 120 hours each week laboratories containing expensive equipment, valuable records, and highly toxic and flammable substances.

Adequate First-aid Fire-fighting Equipment

There are separate approaches to the provision and use of fire-fighting equipment: firstly, where there is mainly reliance on outside help dealing with fires of any seriousness; and secondly, where there is a trained fire crew drawn from the laboratory staff. Differences in the type and range of equipment will occur, as will the procedure. However, the first principles remain constant: the preservation of human life, and the need for rapid action to minimize the effect of fire in all its aspects – injuries to people, and damage to buildings and contents. The first five minutes are vital; what happens in this short space of time will govern the outcome.

Where there is no fire crew the procedure when a fire is discovered has been outlined – evacuation, plus attack on the fire if it is safe to do so. In some countries, including the United Kingdom, training in the use of extinguishers is mandatory. In any case such training is sensible. Excitement, possibly panic, accompanies the discovery of a fire. Such an atmosphere is unlikely to promote a correct response unless training has been undertaken in simulated fire conditions. Many fire brigades have training schools which provide short courses on industrial fire protection, prevention, and fighting. All brigades will give help in simple on-site training. Where there are fire wardens they should preferably undertake the more detailed training and all others should go through the on-site programme. Verbal instruction will not take the place of actually using an extinguisher to put out a fire. A metal tray into which used oil can be poured will provide a flammable liquids fire. When put in a safe place and ignited, suitable extinguishers can be expended in putting it out under the guidance of an instructor. A cube made up from slatted wood sides filled with shavings impregnated with liquid of reasonably low flammability, the whole bound with wire, will provide a fire of carbonaceous material which can be ignited, extinguished, and reignited several times. Experience in actually putting out a fire will bring returns when there is a true fire.

First-aid fire-fighting equipment normally consists of hose reels and extinguishers. Sprinklers and other automatic or manually operated quenching systems are dealt with in the next section. Some hose reels have a control valve fitted to the water supply which must be turned on before water will flow. They are not common now, but a check should be made to identify where this is so. Bearing in mind this reservation, hose reels fall into three main categories. The swinging type is fixed to a wall by a right- or left-hand hinge mounting. There is a controllable nozzle at the end of the 2 cm hose, which has a standard length of 30 m (other lengths are available). The whole assembly is plumbed in to the mains water supply. When 3 m of hose have been pulled out the water supply is automatically turned on.

The fixed type hose reel is substantially the same except that it is permanently fixed flush with a wall. The recessed type is similar to the hinged type but is fitted into a recess. All these reels should be tested regularly by pulling out the required length of hose and directing the water stream through an adjacent window or into a bucket or toilet by operating the controllable nozzle. In this way it can be assured that the hose reel will work when needed. Defects found include water supply interrupted by gravel which had somehow found its way into the reel, and inadequate water pressure. In the second instance pure chance identified this deficiency. In a very large laboratory block hose reels were tested on Saturday mornings. On one occasion due to non-availability of a fireman on Saturday the test was made on a Friday evening. A weak dribble of water came from the hose. Investigation revealed that because of low water pressure the supply company had installed a booster pump to raise pressure in the whole area. This pump operated continuously during normal working hours but intermittently during the evening and night. The remedy applied was to instal an internal booster pump in the boilerhouse of the laboratory complex which started up automatically when the fire alarm operated.

There are certain general rules regarding portable extinguishers:

a) the type(s) of extinguisher should be right for each class of fire which may occur;
b) there should be enough extinguishers in relation to the risk in the laboratory, etc.;
c) the extinguishers should come from a reliable manufacturer and comply with national standards or carry the label of a recognized fire-testing laboratory;
d) they should be serviced at regular intervals, with dates of service shown on an attached label;
e) they should clearly portray the class of fire on which they can safely and effectively be used, and also circumstances in which they must not be used.

Where there is a fire crew mobile first-aid equipment of greater capacity will supplement the fixed location extinguishers. This equipment is likely to consist of wheeled, dry chemical powder extinguishers having a capacity of up to 100 Kg. The discharge throw is up to 12 m at a rate of 2 Kg s^{-1}, and considerable knock-down power is provided for about 50 s. Bearing in mind that other types of extinguisher are available, probably with water from hose reels for quenching effect, this kind of contrivance is most suitable for laboratories. Fire crews should receive adequate training in the use of all equipment on the site, and their role as first-aid fire-fighters made clear. Their task is to attack and contain a fire when it is possible to do so, then to hand over to the professionals when they arrive. This is of course secondary to their search and rescue role should anyone be reported missing. Members of crews need proper equipment, including breathing apparatus and protective clothing of the Heatshield type, which they will don as necessary and in accordance with circumstances. Experience has shown that in the larger laboratory groups there is no shortage of volunteers to form a fire crew and undergo training.

Prevention of Fire Spread

Once a fire is discovered it is of prime importance to confine it to the smallest area possible. This is achieved by construction detail, automatic extinguishment, and safe practices. Each has a part to play but containment is best attained by a combination of all three. Where automatic extinguishment is impractical or difficult, then the other factors will still have an important effect on checking fire spread. The use of fire resistant materials of construction, separation or compartmentation by barrier or fire walls, the closure of stairways, the installation of fire and smoke-stop doors with a given period of fire resistance, will all play their part. Their effectiveness is nullified by improper practices. Fire resistant doors left open, sometimes even chocked open, serve no purpose at all. Doors of this kind held permanently open by electro-magnets linked to the fire alarm system but self-closing when the alarm operates are acceptable, and even desirable, in laboratory connecting corridors which see a constant flow of people and materials. Even these are sometimes reduced in effectiveness by obstructions placed that prevent proper closure. Fire extinguishers have been observed impeding these doors. Wired (or Georgian) glass will prevent the horizontal spread of fire where there are windows in walls between laboratories. It should also be used in doors where visibility allied to fire-stop is required. Flammable liquid stores are dealt with in Chapter 11.

Fixed systems should be tailored to the needs of the area to be protected. Automatic sprinklers are the most extensively used fixed fire-extinguishing system. Insurers allow substantial discounts on fire insurance premiums for buildings protected by sprinklers, and in some countries like New Zealand and the United States they are mandatory in certain circumstances. Water spray is effective on all types of fires where there is no hazardous chemical reaction between the water and the substance that is burning. Generally considered by fire protection engineers as the most effective type of fire-fighting equipment, sprinklers can be dry, wet, and alternate. Wet and alternate are commonly found in the UK but dry systems because of climatic considerations, the construction and contents of the building, and the likelihood of maloperation of sprinkler heads, are conventional in some countries. Wet systems have water constantly in the pipes. Dry have air in the pipes from the main valve groups to the sprinkler heads and usually connected to the town mains. There may be a "dry riser" which is charged by the fire brigade, connecting the rise to the mains. Alternate systems are wet in summer and dry in winter in areas where they may be susceptible to frost. Dry and alternate systems are normally fitted with an air accelerator which facilitates the rapid evacuation of air when a sprinkler head actuates. The sprinkler heads may be of the deluge or spray type. Automatic operation is triggered off by a rise in temperature and a carefully selected system in the light of the protection needed will provide ever-present fire extinguishment.

Local or flood-type carbon dioxide systems are used in laboratories where flammable liquid or gas processes are carried on, in electrical substations, and other locations where fire can be extinguished by diluting the oxygen content of the air or where water must not be used for reasons already given. The carbon dioxide is stored in compressed gas cylinders and released by the operation of a manual device, or automatically via a fire detection system, and is ejected through nozzles close to the expected source of fire, or strategically placed to flood the protected area. Unlike water, foam or other chemical extinguishants it does not usually damage the substances in use, the equipment or the structure itself, although thermal shock due to a rapid decrease in temperature is possible. In areas where people do not work a total flood system is effective and safe provided the location is thoroughly ventilated after a fire has been extinguished. A standard carbon dioxide warning sign should be attached to the door. Where people are likely to be working in an area protected by an automatic carbon dioxide flooding system audible alarms should be linked to the system to alert persons working in the area. A short time lag between the operation of the alarm and discharge of carbon dioxide is advisable in certain circumstances to allow time for evacuation. Not every location is suitable for this kind of system. One particular laboratory working mainly on SPB[2] (low-grade petrol) was protected by carbon dioxide. On the outside wall were large windows. A flash fire occurred, the glass fractured, and the system failed to control the fire. The most effective and rapid extinguishment is attained in small buildings, compartments, laboratories where permanent openings can be automatically shut when the gas is released, and/or having windows of reasonably small area protected by wired glass. Nozzles placed at small localized risk areas can also be effective.

Dry chemical piped systems provide quick extinguishment through an agent which is not toxic or a conductor of electricity and, unlike water sprinkler systems, will not freeze. Inert gases such as nitrogen can be used to reduce the amount of oxygen in air from the normal 21 to 16 per cent or below, depending on the type of combustible material present. Computer installations are fairly common in laboratories and there is a division of opinions as to the best in-built fire protection for them. Halon spheres have their adherents. Carbon dioxide was once commonly used, and still is to a lesser extent. The disadvantages it has are possible thermal shock to delicate equipment and condensation. Water sprinkler systems are common and acceptable. Because an oxygen deficient atmosphere is produced by most systems suitable precautions concerning the protection of personnel are required. All the systems mentioned are effective when properly selected and, most important, regularly maintained and tested.

Minimization of Fire Damage

Many things already mentioned will help to make fires less likely to start, provide rapid extinguishment when they do occur, and help to contain them. All these items help to minimize fire loss. Careless methods of extinguishment and water damage can add substantially to losses suffered through fire. Foam, for example, is an excellent medium for controlling flammable liquids fires, but used indiscriminately it can cause considerable mess. Water damage can be limited by having a supply of sandbags on a pallet adjacent to the building. Placed in doorways or other strategic positions they will contain water within a limited area. Electrical distribution boards are susceptible to water damage. Care in siting them, ensuring they are not placed beneath, for example, an expanded metal floor or other opening, and providing some protection over them, will reduce the risk. When fire inspections are carried out inspecting teams could devote a few minutes to considering the path of water used in fire-fighting and what can be done to channel or divert it from delicate and expensive equipment.

Finally, there has been no attempt in this chapter to present a comprehensive guide to fire-fighting in all its ramifications. Many volumes would be needed to achieve that. The objective has been to set down information which will be of help to laboratory workers in general, and section leaders and managers in particular, in reducing the fire risk at their place of work.

REFERENCES

1. *Journal of the Fire Protection Association*, **112**, "Fire prevention—a guide for management", January 1976.
2. In the UK prior to 1971 fire precautions, including the fire protection of buildings, details of construction etc. were dealt with by the Factories Act, 1961 and the Offices, Shops and Railways Premises Act, 1963 supplemented by by-laws made by local authorities. Because legislation was fragmented, the Fire Precau-

tions Act, 1971 dealing solely with fire precautions was passed. The Health and Safety at Work Act, 1974, further strengthened the 1971 Act. Most other countries have legislation dealing with fire precautions, fire resistance of structures, etc. Canada has its Fire Marshals Act, there are regulations in Czechoslovakia, Netherlands, USA, West Germany and most other countries. Details of most can be found in "The Law and Practice Relating to Safety in Factories, Part II, Legislation" published by the International Labour Organization.

FURTHER READING

Auto-ignition Temperature of Organic Chemicals, *Chemical Engineering*, **75–80.**
Code of practice for the keeping of liquified petroleum gas in cylinders and similar containers, HMSO, London.
Code of practice for the recovery of solvents, Chemical Recovery Association, London.
Dangerous substances, guidance on dealing with fires and spillages, HMSO, London.
Fire-fighting Equipment, Fire Alarms and Fire Drills in Offices and Shops, Safety, Health and Welfare Series No. 5, HMSO, London.
Flammability Characteristics of Combustible Gases and Vapours, US Bureau of Mines, Bulletin 627, Washington DC.
Flash Points, The British Drug Houses Ltd., Poole, UK.
Freeman, N. T. and Thacker, B. W. (1979). "Appointment with Fire", Alan Osborne Associates (Books) Ltd, London.
Industrial Solvents, Flammable Liquids and Low Melting Point Solids, Fire Protection Association, Aldermary House, Queen Street, London EC4.
Marsden, C. and Mann, S. (1963). "Solvents Guide", Macmillan, London.
Wilbraham, A. C. (1979). Fire Safety and Fire Control in the Chemical Laboratory, *Journal Chemical Education* **56,** Part 10, A311–A315.

10

Safety Procedures and Documentation

In a number of countries emergency procedures, properly documented, regularly practiced and approved by the appropriate enforcing authority, are mandatory in high risk locations. Often there is a link between quantities of hazardous materials processed and stored, and legislative control. Because in normal circumstances these quantities are not found in even the largest laboratory complexes there is a reluctance to set up emergency procedures in them other than those concerned with fire. This is wrong. Although toxic and flammable substances may be handled and stored in relatively small quantities, their variety and complexity is normally greater than in manufacturing areas. Additionally, before large-scale production begins the research and development work which has preceded it has reduced or eliminated hazards, a factor which applies in a more limited way while laboratory activities are in progress. Where laboratories are incorporated in a manufacturing facility they will be included in the site emergency procedure. The important consideration is that there is such a plan in existence irrespective of size.

Careful preplanning is necessary if an effective scheme is to be operated. It is of little use to think of the introduction of emergency procedures when disaster is imminent. An important preliminary is to gain acceptance of the need. Laboratory people may well consider they are not likely to be involved in a disaster or near-disaster. Those working in laboratories on a plant where ammonia was in use on a fairly large scale probably came into this category until a tank being filled with ammonia was overfilled. The relief valve operated spraying liquid ammonia into the atmosphere. The fumes were drawn into the air-conditioning system of the laboratories, driving everyone from the building. In the uncontrolled rush for the exits one employee ran blindly into the road and narrowly escaped being struck by an approaching car. Action was subsequently taken to minimize the possibility of a recurrence by installing tank level controls and scrubbers in the intake side of the air-conditioning, coupled with improved emergency arrangements in the laboratories. Had these arrangements been made before the incident, those who suffered from ammonia inhalation would have escaped this unpleasant experience. In another case an explosion in a laboratory which formed part of a large main research building was followed by fire, a lesser secondary explosion, the production of noxious fumes and smoke, and the death of both people working in the laboratory. One died of shock due to loss of blood and chemical-thermal burns, the other following the inhalation of toxic fume. Damage extended beyond the laboratory in which the first explosion occurred. One of the chemists had been working on the synthesis and characterization of ethylene dioxyamineperchlorate (EDAP), a sensitive material which, it is believed, was being purified at the time of the explosion.[1] There are other examples, but the two quoted should help to convince laboratory workers of the need to devise an emergency scheme where none already exists. It is again emphasized that the detail which follows can be scaled down to meet the needs of the smallest installation.

Every employer has a moral obligation to safeguard the well-being of his employees, those who live in the vicinity of the enterprise, and others who may casually be in the area. The report of a disaster at Flixborough, in the United Kingdom[2] stated: "In any area where there is a major disaster hazard a disaster plan for the co-ordination of rescue, fire-fighting, police and medical services . . . is desirable". From the purely materialistic angle a corporate body like a company owes a duty to share or stockholders to do everything possible to ensure that nothing is neglected that would safeguard the assets of the company. Where an individual is the employer he has a duty to himself and his dependants to do the same. It has been argued that knowing an emergency scheme exists may create alarm and despondency among people living near a laboratory installation. This can be countered by the consideration that although every known precaution may have been taken in the design, construction, and operation of laboratories, accidents may still occur. Techniques are well established for dealing with the majority of these. Surely it must be reassuring to nearby residents to know that if a large-scale incident should occur those responsible for controlling the laboratories have well prepared and rehearsed plans for dealing with the situation?

What about size and different levels of risk? Some laboratory work is clearly more hazardous than others, but the need for being prepared for an emergency has little connection with size or scale of risk. It would be a brave laboratory manager who could say with complete confidence that the laboratory(ies) under control was so small, and the work of such a nature, that nothing would happen within the confines which could present a hazard to the surrounding community as well as to those working within it. A properly conceived emergency plan will take account of four phases:

Phase 1: Preliminary Action

A plan should be prepared tailored to meet the special requirements of the site and buildings, the work carried on, the quantity and nature of substances stored, and the surroundings. Every employee should be familiarized with the details of the plan. The purchasing and positioning of essential equipment should be done. The training of personnel involved is needed. There should be an initiation of a programme of inspection of potentially hazardous areas, a testing of warning systems, and evacuation procedures, laying down specific periods at which the plan should be re-examined and updated.

Phase 2: Action when Emergency is Imminent

There may be preliminary warning of an emergency. Accidental damage to apparatus or a suspected leak of highly flammable or noxious substances may initially create a potentially dangerous situation. Disastrous floods or hurricanes are often preceded by warnings. The warning period should be used to assemble key personnel, review the standing arrangements to consider if any modifications are necessary in view of the nature of the pending emergency, give advance warning to external authorities, and test all systems connected with the emergency scheme.

Phase 3: Action During the Emergency

If Phase 1 has been properly carried out (and Phase 2 where applicable) this action proceeds according to plan. However, it is likely that unexpected variations in predicted emergencies may occur. The decision-making personnel, who will have been chosen beforehand with this in mind, must be able to make incisive and rapid judgements and see that proper action follows the decisions made.

Phase 4: Ending the Emergency

There must be a procedure for declaring laboratories or areas safe, an orderly reoccupation of buildings where this is possible, and action detailed for rehabilitation.

Liaison with Emergency Services and Others

Because the emergency services are going to be concerned in the operation of the scheme, their representatives must be involved at the planning stage.

When a rough draft has been prepared a meeting should be called attended by delegates from the laboratories, the police, and the fire authority. Dependent on circumstances, others invited could be from neighbouring enterprises because the draft scheme to be discussed might conflict with others in the area; it might reveal the need for integration with existing schemes; it might even inspire others to initiate a plan. Mutual aid arrangements sometimes evolve from such a meeting, incorporating the pooling of on-site emergency equipment, the development of common training plans, and the establishment of intercommunication systems. The specific contributions sought from the authorities are an appreciation of the nature of the risks, planning and ensuring the adequacy of equipment, considering the need and methods for achieving road blocks, traffic diversions, warning the public and evacuation of surrounding areas, the possible type and number of casualties, and the availability of specialist treatment which may be required. All those present will be concerned in the way in which an emergency is to be declared, and how they will be alerted.

Details of the Scheme

There must first be a definition of what constitutes a major emergency. A suggestion is "a situation that may affect several departments within a laboratory complex and/or endanger the surrounding community". It may be precipitated by a malfunction of normal laboratory procedures, by the intervention of some outside agency such as a crashed aircraft, storm or hurricane, flooding, a deliberate act of arson or sabotage, or a civil/military disturbance. Emergency conditions most likely to arise within laboratories are:

1. a large or rapidly escalating fire, or explosion;
2. a large-scale release of toxic substances;
3. a release of radioactive substances;
4. any combination of these conditions.

Laboratories should be examined to indicate the likelihood of any of these conditions occurring through internal causes. An appraisal of accident reports and recorded "near-misses" can provide useful information.

When considering the type of emergency which may occur, account must then be taken of the areas likely to be affected. The interdependence and proximity of laboratories, storage areas, services, and other buildings should be considered. Particular attention should be paid to the effect of wind direction and strength, and the effect these may have on the spread of fire and toxic substances. The civil authorities will need information about areas outside the laboratories which could be affected by toxic fume. A simple addition to an ordinary anemometer dial will provide this information, as shown in Fig. 10.1.

Raising the Alarm

Apart from rehearsals and periodic drills on a scale and frequency related to the degree of risk, the special procedures for handling major emergencies must only be initiated when such an emergency exists or is imminent. A limited number of designated senior personnel should be assigned the responsibility of deciding if such a situation has arisen or is arising, and only those personnel should have the authority to implement the procedures. When selecting authorized personnel, thought must be given to their availability, particularly outside normal working hours and at weekends. They should be chosen also because of their ability to make rapid and accurate assessments of emergency conditions.

Types of Alarm

The choice of an internal alarm system will be governed by local conditions

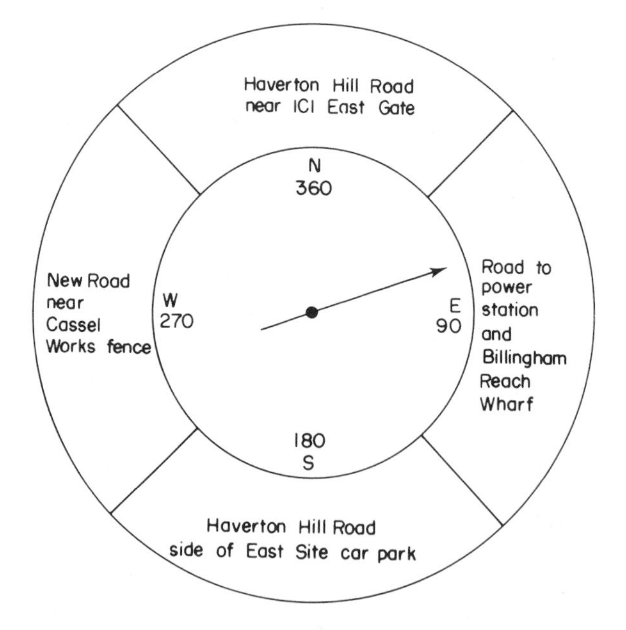

Fig. 10.1. An anemometer in use at one of the sites operated by Tioxide International. The inner circle shows wind direction. The indicator arrow at the same time shows in the outer circle areas to which people can safely be evacuated.

such as size, the work carried out with special regard to the interdependence of laboratories and buildings, the rapidity with which escalation can occur and the existence of alarms already in use to signal other emergencies. The alarm note for a major emergency should be easily distinguished from that used for fires and other incidents not within the definition of a major emergency.

For large laboratory complexes, it is preferable that the alarm be selective, i.e. capable of being sounded in selected areas as well as over the whole site. Unnecessary shutdown of laboratories in which work is in progress that would not be affected by the emergency can then be avoided. In some locations, a two-stage warning can be used. The first stage acts as a warning to all areas that a major emergency is imminent and the second selective stage, which may simply be the first stage alarm re-sounded for a longer period, confirms in the areas likely to be affected that the major emergency exists. In other cases, where the nature of the operation carries a high risk, the use of a single alarm system may be preferable.

Emergency Controller

Senior members of staff with a thorough knowledge of the work carried on and its associated hazards should be named as Emergency Controllers. Account must be taken of their availability (e.g. their normal work should not involve them in extensive travel) and of the personal qualities required for the post. They must be clear-headed in times of panic, decisive and capable of leading under the worst of conditions. Out of normal working hours, the senior member of management on site should take initial control until relieved by the duty Emergency Controller.

The Emergency Controller should wear a distinctive garment such as a brightly coloured or luminous jacket and helmet to make him easily recognizable. A deputy should also be appointed to take charge when the Emergency Controller is not available, i.e. during holiday periods and sick leave.

A typical list of duties for the person in control is:

1. Occupy the Emergency Control Centre.
2. Put on distinguishing clothing.
3. Check that the warning and call-out routine is being followed.
4. Check that the laboratory fire crew (if any) is in action, and that the casualty and personnel-checking routine is working.
5. Establish liaison with senior police and fire officers when they arrive, and arrange with them the setting up of alternative communications systems. Advise on specific risks and how they should be tackled.

6. Determine action necessary to deal with changing circumstances.
7. Decide when emergency is over and what is to be done about reoccupation, salvage operations, rehabilitation, and so on.

Emergency Control Centre

A good communications system is essential. A Control Centre should be established and equipped with adequate means of receiving information from the forward control and assembly points, transmitting calls for assistance to external authorities, calling in essential personnel and transmitting information and instructions to personnel within the site. Alternative means of communication must be available in case the main system becomes inoperative. If all systems fail, members of the emergency team should report to the Control Centre to act as runners.

The Control Centre should be sited in an area of minimum risk, so far as is possible, and where the maximum use can be made of existing communication systems. An alternative centre should be provided at a different geographical location in case the main centre becomes inoperative or uninhabitable. Centres should be sited near good roads so that transport can gain ready access, and in order to facilitate the use of a radio-equipped vehicle if other communication systems fail. Each Control Centre should also contain:

1. A map of the site showing:
   a) means of access to the affected areas;
   b) access which may be temporarily unusable, e.g. because of repair work;
   c) locations of emergency equipment such as breathing apparatus, fire hydrants, protective clothing, medical supplies, and neutralizing material.
2. A map of the surrounding district covering a circular area with a radius of one mile, with the laboratories as the centre. This map should be equipped with a device which will indicate the area likely to be affected by wind-blown materials. An example is shown in Fig. 10.2.
3. A list of materials used on the site, their properties and locations. Special risks, e.g. radioactive materials, should be conspicuously marked on the map. Copies of the list should be available for issue to key personnel and emergency crews.
4. Protective equipment including, where necessary, impervious clothing and breathing apparatus preferably equipped with a voice amplification device.
5. A list of the names, addresses, internal and home telephone numbers of key personnel, and a second list having details of *all* personnel.

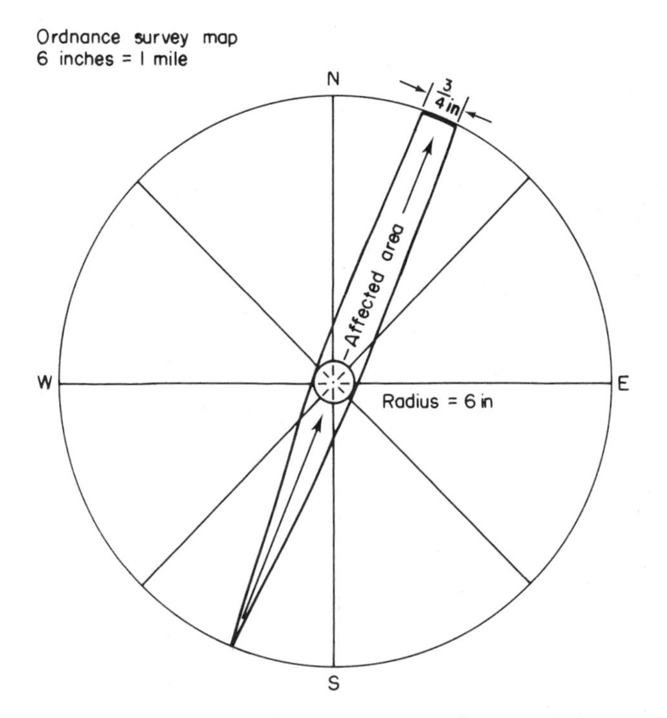

Fig. 10.2. A circle, radius 6 in, with the laboratory location as centre, is drawn on an ordnance survey map, 6 in = 1 mile. A transparent plastic strip 12 in long ($\frac{3}{4}$ in wide for 6 in, reducing to a point thereafter) is pinned at its centre to the centre of the circle. When the pointed end of the strip is placed on the direction from which the wind is blowing, the broad half indicates the area most likely to be affected by wind-blown materials.

If possible, a detailed log of events during the time of an emergency should be kept. This will be useful when a review of the emergency procedures is carried out in the light of the experience gained.

Transmitting the Alarm

Apart from prearranged practices, the procedures for major emergencies must be initiated only when an emergency exists. It is important to establish with each authority a suitable method by which warning of a major emergency can be transmitted to them. Normally, this can be achieved by a predetermined short message, transmitted via an emergency line or by the normal GPO lines. The warning message should include advice on routes to the laboratories which may become impassable. Alternative routes can then be used.

Call-outs

A list of employees needed in a major emergency should be drawn up, together with their internal and home telephone numbers and addresses. The list should be in the Control Centres and constantly updated to take account of changes in personnel, telephone numbers, and addresses. The list should be divided into two parts:

1. Those who are required immediately, e.g. certain section managers, engineers, specialists, and chemists.
2. Those who may be required at a later time, e.g. personnel department, catering staff, etc. These can be called in as required, without delaying other more urgent calls.

On-the-spot Action

A clear division of responsibility is needed. The immediate location of the emergency will be dealt with by vigorous action by supervisors and other personnel to close down and make safe areas which are affected. The Emergency Controller will decide how much can safely be done. It may be possible to minimize the effects of the emergency, e.g. by isolating fractured pipelines from which flammable or toxic gases may be escaping.

The preservation of human life and the protection of property is the responsibility of the laboratory fire crew. They will handle casualties, conduct evacuated personnel to the appropriate assembly points, and fight fires which may occur, as directed by the Emergency Controller.

Seriously injured persons must be taken to hospital with the least possible delay. Temporary first-aid facilities may be needed at points in a safe area accessible to ambulances.

Evacuation

In the immediately affected area, it will almost certainly be necessary to evacuate non-essential personnel. There will need to be an evacuation alarm which, as with the major emergency alarm, should be distinctive, known to everyone, and selective. It will be sounded on the instruction of the Emergency Controller. Assembly areas, to which evacuated personnel will report, should be chosen beforehand with due regard to wind direction. Each area should be clearly marked and known to all employees. It should be in the control of an appointed member of staff and there should be a means of communicating with the Emergency Control Centre. Although it is not always possible to do this, escape routes should be planned so that people will not have to pass through danger areas.

Accounting for Personnel

A check is necessary to find out if people are missing, or that everyone can be accounted for. This can be achieved by taking a roll-call at each assembly point and checking names against a nominal roll of employees. However, there are considerable difficulties. An up-to-date nominal roll is extremely difficult to maintain. For most of the time it will not be completely accurate. Some employees will have left the company, others will have joined, and there will be absentees. Other employees can usually be relied upon to provide the names of such people. Contractors' employees and visitors who may be on the site must be accounted for. It is up to each employer to devise a scheme which is practical and workable in the light of its own circumstances. A suggestion as to how to deal with situations in which flexitime operates is in Chapter 9—Fire Protection and Prevention.

Access to Records

Relatives of injuried or dead employees will need to be informed. This duty is normally undertaken by the police. It has already been suggested that at each Emergency Control Centre there should be available a list of the names and addresses of all employees.

External Communications

Arrangements should be made to ensure that urgent calls can be transmitted from the laboratories without delay. The use of Police or Fire Authority radio vehicles offers a convenient standby in case the telephone system is damaged. Even where the telephone system remains in use, difficulties may arise if incoming calls from anxious relatives and people volunteering assistance, jam the board. At least one ex-directory telephone which bypasses the switchboard will ensure that outgoing calls can be made. It is also possible to provide lines on the switchboard which can be used for outgoing calls only. The telephone company or post office exchange may be able to arrange for non-essential calls to be handled by them. Instructions to employees can include asking for their co-operation to persuade relatives not to try to contact the laboratories in the event of an emergency and, also, not to telephone in themselves—they will be sent for if needed.

Public Relations

Inevitably, a major incident will attract the attention of the press, television, and radio services. It is essential to make arrangements for official releases of information to them. This should be handled by a public relations officer or by a member of staff not otherwise engaged in handling the incident. All other employees should be instructed that it is of the utmost importance not to release information but to refer enquirers to the appropriate appointed person who will be the only one with accurate information concerning the overall situation. In the interests of their own safety, newsmen, photographers and others should not be allowed free access to the laboratories, though every effort should be made to provide them with accurate information if only to avoid misleading, and perhaps exaggerated, accounts of the incident being published.

Catering

Emergency teams will need to be provided with refreshment if the incident lasts a long time. Arrangements for calling out catering staff are therefore necessary. If the on-site facilities cannot be used, it may be possible to arrange for the use of local authority or neighbouring company facilities.

Training

It is difficult to predict how people will react in emergencies. Knowing what to do and practice in doing it will increase the probability of the right thing being done at the right time. Detailed written instructions should be prepared and issued to all those required to take action to control or minimize the effects of a major incident. Experience has shown that some of this detail tends to be forgotten when an emergency occurs. For this reason it is useful to prepare very short summaries of the actions to be taken to serve as an aide-memoire when an emergency occurs. A simple flip chart will achieve this.

Thorough training is important and should include practical exercises which are as realistic as can be achieved. These may be held at times of planned shut-downs or at some other convenient time so as to avoid undue interference with work in progress. Where this is not possible every effort should be made to involve as many personnel as possible. Information about the major emergency scheme should be passed on to new employees on their first day at work. Exercises should include the participation of those outside services that will be called upon. Where mutual aid schemes with neighbours are in existence, all the participants should take part. Fire authorities in particular should be invited to send firemen to laboratories so that they can get to know the layout, procedures, and special hazards which they may encounter.

Post-emergency Action

If Phase 4 has been correctly planned, detail will be available of demolition contractors, construction companies, suppliers able quickly to provide replacements for damaged or destroyed equipment, even estate agents who can quickly negotiate the rental of temporary premises if required.

It may be thought that some of the suggestions made go too far, particularly in the area of post-emergency planning. This is not so. The time to plan is in the sober atmosphere of normal operation, not when an emergency has occurred and the adrenalin has been ejected from the ductless glands into the bloodstream. It is in any case almost always too late to take effective action when disaster is upon us.

What follows has already been said, but is re-emphasized because of its importance. Because the detail given relates to the larger company, small operators should not therefore be deterred. All that is needed is a scaling down of the suggestions made to fit the size and resources of the laboratories concerned. The expertise available from experienced people in this field may with advantage be used. This is not to imply that a "do-it-yourself" job will not be effective.

Finally, the whole matter of emergency planning is put in perspective if one subjects oneself to a simple test. If a catastrophe occurred in laboratories for which you had managerial responsibility, how would you answer these questions should you find yourself in the witness box at a coroner's court, or a court of inquiry?

1. Did you have a major emergency scheme?
2. Did the people for whom you had a responsibility know what to do if an emergency occurred?
3. Were contractors' employees and other invitees on the site aware of what to do in an emergency?
4. If you had an emergency scheme, was it regularly practised and in efficient working order?

Should there be an unsatisfactory answer to any of these, then managers should put their house in order as soon as possible. Otherwise they are failing in their statutory and moral duty to those who respond to them, those living or working around them, and their dependants.

PERMIT-TO-WORK SYSTEMS

These systems are common in production areas, less common in laboratories. Considering the experimental and often hazardous work carried on in them from time to time, this is an unacceptable situation. It is unrealistic to suggest that work carried on in a workshop or factory should be subject to a permit to work while the same or similar work in laboratories is not. A simple example is cleaning or maintenance of electrical equipment. In one case observed in an otherwise efficiently run and well-organized laboratory block, two electricians were putting in an additional circuit for the pathological laboratory. They had isolated the appropriate section of the distribution board by removing fuses which they had then placed in the bottom of the fuse box. This was the state when a laboratory assistant needed one of the isolated circuits so he replaced the fuses. Fortunately he was seen to do this by someone aware of the overall situation and the fuses were quickly removed again. But for this fortunate intervention the electricians temporarily absent on a meal break would have unwittingly restarted work on live equipment with disastrous results. In the production areas adjacent to the laboratories this work would have been subject to a permit which would have specified proper isolation by switching off and then locking off using a multi-hasp. Each electrician would have placed his padlock on the hasp, retaining the key in his pocket. No-one except the two electricians in unison could then re-energize the circuits. After this incident the system was extended to the laboratories. The lesson learned could have been driven home by fatal injuries but for the intervention mentioned.

A permit-to-work system will not of itself make work absolutely safe because the complete elimination of the human element is unachievable. If a system is developed in which the irreducible minimum of human involvement is attained then it will go a considerable distance along the path to absolute safety. When a permit system is first introduced resistance is sometimes encountered. Those called on to sign related documentation may be reluctant to do so. A training programme is therefore an essential accompaniment to its introduction. Programmed learning has produced excellent results in this context. In one instance new employees with no previous knowledge of a permit-to-work system scored an average 91 per cent in the post-programme test. The same group retested six months later showed a retention level of 86 per cent. Retraining sessions (an important element of any system) produced an average test score of 95 per cent. It might be thought that an important element of accident avoidance such as this should not be acceptable unless those involved showed an absolute knowledge in tests following training. In fact the tests covered all aspects of the system, its operation, and the duties of everyone involved. The trainers had to be satisfied that each individual had an adequate knowledge of his or her personal involvement, and until this goal was achieved training was repeated. Below is a general outline of aspects of permit-to-work procedures.

From time to time it becomes necessary for personnel to carry out work which could give rise to hazardous situations in the following events:

a) The presence, in the first instance, of inherent danger, e.g. flammable or toxic substances, moving parts of equipment, electricity, etc.
b) The introduction of dangerous acts or conditions after work has commenced on jobs previously thought to be safe.

In laboratories a careful assessment must be made of work to be undertaken and a decision made as to the need for a Permit to Work. In many cases, people who are knowledgeable about their own particular operations rely largely upon word of mouth to relay instructions, but even in the best-run establishments and with the very best of personnel involved, forgetfulness, sudden emergency conditions arising elsewhere which divert attention from the job in hand, illness of the person who has verbally authorized work etc. can lead to a loss of direct control which, in turn, may lead to accidents. The United Kingdom's HM Factory Inspectorate summed up the situation in this way:

> Verbal instructions, requests or promises are always liable to be misheard, misinterpreted or forgotten and must never be regarded as a satisfactory basis for action on which men's lives may depend.
>
> Experience has shown that to achieve the maximum degree of safety, the "human element" must be eliminated as far as possible by using a system which requires formal action. Such a system is that of "Permit to Work"...

Purposes of Permits to Work

1. To prevent the interaction and interfaces of dissimilar intentions and operations, i.e. "blocks" other work.
2. To provide a check list for things to be done to achieve a state of safe working before work commences.
3. To provide some measure of work progress.
4. To give written notification of work completion.
5. To provide control.

Documentation

Whilst there can be many variations on the actual format of the Permit-to-Work document, all should comply with the following criteria and include:

1. Precise, detailed, and accurate information.
2. The time of coming into force.
3. The time of expiration.
4. No variations unless the Permit is cancelled and a new one issued.
5. Can only be cancelled by issuing authority.

6. Issuing authority must be satisfied that all specified actions have been taken.
7. It must be the master instruction overriding all others.
8. Person who accepts Permit is responsible for compliance.
9. Copy displayed at site of work.
10. Only the work specified to be done.
11. Signed cancellation when work completed.

Information Required

a) Accurate identification of the part of building or laboratory which is to be worked on.
b) Precise and detailed description of all the work to be done.
c) The times between which the Permit will be in force.
d) The name of the person(s) to whom the Permit relates.
e) Complete description of the precautions to be taken to ensure integrity of the systems and equipment to be used. It may be necessary to supplement the Permit with a Clearance Certificate for such things as atmospheric tests or electrical isolation. The person who carries these out must sign the Permit in the appropriate place.
f) Statement of any special equipment required.
g) Details of any special precautions necessary such as the level or degree of supervision required, or the type and frequency of atmospheric monitoring.

Authorization

The issuing authority must be a person who has autonomous control over all matters which can directly or indirectly affect the work in hand. He must satisfy himself that all necessary precautions have been taken, ideally by personal inspection. If he must delegate, his nominee must sign the Permit to the effect that the checks have been carried out.

Acceptance

The person in charge of the work to be done must sign the Permit to acknowledge that:

a) He has read and understood the conditions and precautions.
b) He agrees to comply with the conditions.
c) He has clearly instructed all persons in his charge regarding the precautions to be taken.

Distribution

1. One copy of the Permit to be retained by the issuing authority.
2. One copy of the Permit to be given to the person in charge of the work to be done (whenever possible, this copy to be displayed at the place of work).
3. Copies of any Clearance Certificates to be attached to the Permits. One copy of the Clearance Certificate to be retained by the issuer.

Cancellation

After work is completed the person in charge must sign to state that his/her personnel are no longer involved, and that any specialist equipment, substances etc. have been removed or made safe. The issuing authority must sign to revoke the Permit when work is completed or if changing circumstances render the conditions and precautions of the existing Permit inoperable.

Renewals

Whether or not Permits may be renewed on a day-to-day or shift-to-shift basis will depend upon the circumstances of individual operations and locations. Certainly there will be a need to renew Clearance Certificates in respect of atmospheric conditions where there is a possibility of continuous or intermittent release of noxious or flammable substances, or similar hazardous conditions. An example of a Permit-to-Work system in operation when maintenance of X-ray equipment is in progress is shown in Chapter 6.

DOCUMENTATION COVERING EQUIPMENT LEFT UNATTENDED

Reference was made on page 32 to equipment which may be left running overnight, or at some other time outside normal working hours. A special permit is needed in these circumstances. It should set down the laboratory or area concerned, the name of the person using the equipment and their home telephone number, the emergency telephone number to be used if necessary, brief detail of the work in progress, and any special precautions required. The starting and ending times and dates should be shown and the permit signed by an authorized person. One copy should be placed on or near the equipment, the other handed to the security or other personnel who will carry out routine checks.

REFERENCES

1. Case History No. 1622, Volume 3, Case Histories of Accidents in the Chemical Industry, Manufacturing Chemists Association, Washington DC.
2. "The Flixborough Disaster", Report of the Committee of Inquiry, p. 37, HMSO, London.

FURTHER READING

Codes of Practice for Chemicals with Major Hazards. Chemical Industries Association, Albert Embankment, London.
Recommended Procedures for Handling Major Emergencies, ibid.

11

Safety in the Laboratory Offices, Stores, and General Service Areas

The ancilliary buildings or areas essential to the efficient operation of a laboratory complex, or small rooms which form an adjunct to the building's main functions, should be subject to the same health and safety requirements as the laboratories. Examples include canteens, rest rooms, stores, and offices. This is recognized in most countries by a legal code which applies either specifically or generally to, for instance, offices and hazardous storage areas. In the United Kingdom there are the Offices, Shops, and Railway Premises Act 1963, the Highly Flammable Liquids and Liquified Petroleum Gases Regulations 1972, and certain clean food requirements which apply to eating areas.

It is one of the oldest tenets of health and safety that injuries seldom occur because of well-known and defined hazards. In laboratories where cyanide is handled injuries involving this substance are a great rarity. People working with it are conscious of the associated hazards and avoid them. But the converse also applies: if work is carried on in what is considered to be a safe area there is a relaxation of care and attention so that avoidable accidents occur. Offices are relatively safe places of work yet 5000 disabling injuries happen in them in the United Kingdom every year. Many more times this number of minor injuries occur. Occasionally office accidents are accompanied by very serious injuries, a factor surprising to those not normally involved in this field. Any accident is the result of failure to control conditions or circumstances. While never excusable, it is a little more understandable when incidents happen in laboratories where the nature of the work is inherently hazardous than when they occur in the often calmer, more relaxed, and certainly safer surroundings of an office.

Analysis of disabling injuries in offices shows that the main causes are falls (using makeshift means of access to reach files, books, and other things above ground level), handling of materials and equipment (lifting boxes of stationery), falling objects (books and files insecurely placed on top of cupboards), machinery (unskilled repairs to equipment, guillotines), moving furniture around, fire (carelessly discarded smoking materials, wipers impregnated with cleaning solvent thrown into waste baskets), and electricity (unskilled repairs to plugs, overloaded outlets). The examples given are not exclusive. They happen in offices every working day.

GENERAL CARE IN THE OFFICE

Planning

The arrangement of equipment, furniture, and fittings needs careful planning for space to be used to the best advantage. Where possible the eventual user should be consulted, otherwise carefully sited furniture may be moved when the occupier moves in. Free movement should be possible avoiding contact with sharp edges of cabinets, tables or desks, important where metal furniture with sharp edges is present. Equipment with drawers should be sited so that obstruction is not caused when they are open. The location of telephones and electrical cables needs thought so that trailing wires are avoided. Ideally all such wiring should be overhead with easily reached adapters for connecting with equipment. This arrangement is safe, but not popular because it is not aesthetically pleasing. An alternative is to site outlets as near as possible to equipment, then secure cables under purpose-made covers. The heating source needs careful consideration. Air-conditioning is the ideal, recessed steam or electrical radiators acceptable. Gas or electric fires should be sited and used with care. Those with open elements should never be used, and combustible materials should be kept well away from heat sources. Careless disposal of smoking materials has already been mentioned as a frequent cause of fire. Is it feasible and reasonable to impose a "no smoking" rule? If not, a safe area in which smoking is permitted may be provided with self-extinguishing ashtrays and a source for lighting cigarettes that avoids the use of matches. It is reasonable in any case to impose a "no smoking during the last half hour at work" rule.

Tidiness

Carefully planned office arrangements as suggested are reduced in effectiveness by untidiness. There should be adequate storage space for all but work currently in hand. Waste bins should be adequate in size and number, thoughtfully placed, and regularly emptied. Sharp objects should be placed in special containers to avoid injury to cleaners.

Falls and Collisions

Many of these are caused by familiarity, absentmindedness, or distraction through, for instance, talking to other persons. A faulty carpet or spillage of water on the floor may be a contributory factor; such a hazard should be seen and removed. Sensible rules are: move carefully, avoid distraction by gossip (as opposed to essential communication); use handrails on stairs; do not carry objects which obstruct vision; do not run; walk down the centre of corridors to avoid suddenly opened doors; use proper means of access to shelves, etc. Some of these things may be considered far-fetched and unnecessary. They are achievable, and contribute not only to accident avoidance but also to efficient operation.

Lifting and Carrying

Office users are not normally trained in correct methods of lifting and carrying, which is unfortunate. Back injuries are far less likely to happen if these precautions are observed when lifting and moving objects:

1. Is there a real need to lift and move something anyway? Is help needed?
2. Use a trolley if possible.
3. Take a firm grip.
4. Do not swivel the back from the hips.
5. Keep a straight back when lifting, bend knees, then straighten legs to lift.
6. Make sure forward vision is unobstructed when carrying.

Office Equipment

The hazards presented by even sophisticated office equipment are straightforward. When breakdowns occur, call in the experts. Treat electricity with respect: there should be a routine inspection of wiring, plugs, and connections. Sockets should never be overloaded by using multi-point adaptors. A major store fire in Manchester, England, was suspected as having such a cause, and many lives were lost as a result. When office machinery breaks down it should be disconnected and an "out of order" notice placed on it until it is repaired. All power-operated machinery should be switched off when not in use and effectively disconnected from the mains when work ceases each day.

Filing cabinets, particularly the four-drawer type, can be an accident source if the upper drawers only are filled, particularly when pulled open. The load should be spread evenly over all drawers and the upper drawers opened with caution. For extra safety a 2 cm batten can be placed beneath the bottom front. Some copying machines use chemicals: the instructions regarding these should be read and specified precautions followed explicitly. Chemicals used though not particularly corrosive can cause skin irritation and any spilt on the skin should be thoroughly washed off immediately.

VISUAL DISPLAY UNITS

The small television screens of Visual Display Units (VDUs) show information in words and figures either derived from, or fed into, a computer by operating a keyboard. They are widely used in offices for such tasks as stock control, airline seat reservations, bank card accounts, medical records, and scientific research. Rapid increase in their use has resulted in speculation about effects on operators' eyesight.

It is claimed by Rosenthal and Grundy[1] that VDUs do not cause damage to the eyes or aggravate any deterioration in the operators' sight but, because of the concentrated nature of the task, the operator may be made more aware of any existing visual deficiency. This awareness can take the form of eyestrain which can manifest itself as tiredness, irritation, headaches, screwing up of the eyes, and photophobia. Precautions can be taken to alleviate these symptoms and these may be summarized as follows below.

VDU Equipment and its Maintenance

A unit should be selected which gives optimum clarity of screen presentation. Important factors are size and spacing of characters, focus, brilliance, contrast, colour, layout, and flicker. The equipment should be regularly maintained to ensure that the quality of presentation has not deteriorated. It has been contended that a yellow/green combination of colour is most suitable for the normal eye but the choice of colour is a personal one and a variety of colour combinations is available. Controls for focus, brilliance, and contrast should be readily available to operators so that they can adjust them to their liking.

Siting the Equipment

The correct positioning of the equipment is particularly important and it will vary according to the nature of the eyesight of the operator. Keyboards are normally movable and can be readily adjusted to a position comfortable to the operator. If the operator is working from documents then the position of these should also be adjustable. Working distances are usually within the range 33–100 cm and the positions of the keyboard, screen, and documents must be arranged to be within these distances. Young persons with normal eyesight can readily focus over this range but, as persons age, their ability to do so decreases and the need to wear spectacles increases. There has been much discussion on the preferred type of spectacles, particularly of the bifocal type, and the VDU operators should inform opticians of the nature of their work when their eyesight is being examined.

It is recommended that VDU operators should undergo an eyesight test before starting work so that any deficiency can be remedied. Colour blindness is not necessarily a bar to being a VDU operator provided an effective contrast between letters and screen can be obtained.

Lighting

The equipment should be sited so that glare and screen reflection are eliminated. Some reduction in the overall lighting of the office may be necessary and levels of 100–300 lux have been suggested to be suitable. Additional local lighting for documents may be necessary but this must be carefully screened from the operator to reduce the contrast between the light and the screen. Lighting is very much an individual matter and the operators should be able to adjust the standard of illumination to their own requirements.

General

Persons suffering from epilepsy, acute migraine, or nystagmus should not be selected for work with VDUs. Environmental matters such as temperature, humidity, noise, and decor tend to assume a greater importance than with the normal office worker and should therefore be given even greater than usual attention.

The possibility of radiation being emitted by the screen has been investigated and shown to be non-existent in the light of the knowledge available. Attempts have been made to lay down maximum periods of operation per day but as there are manifest differences in the characteristics of operators, wide variations are inevitable. In general, a break from work should be taken if eye strain symptoms are experienced.

LABORATORY STORES

In laboratory blocks of reasonable size specific rooms or buildings are allocated for the storage of chemicals and apparatus. In smaller locations an area may be set aside for this purpose. Supervision of this store may vary from part-time duty if the value and amount of material are small,to full-time employment for a team of store people where the operation is large. In all cases the person in charge should be competent to discharge the control function and have an adequate knowledge of the materials stored. Those working in the store should be aware of the dangers of the materials and substances handled. Entrance should be restricted to authorized personnel.

Certain basic rules should be followed. Chemicals should be stored in a separate area from equipment on strong shelving fitted with a lip to prevent bottles being inadvertently pushed off. Corrosive chemicals should not be stored above waist height and should be in non-corrodible trays which will retain the chemical if the containers are broken. Chemicals should not be stored together, enabling mixing to occur in each tray, e.g. concentrated sulphuric acid mixed with ammonia liquor would produce a very hazardous situation. All chemicals should be correctly labelled and the labels should be designed to be clearly distinguishable. A serious incident occurred in the authors' laboratories when concentrated hydrochloric acid was mixed with concentrated sulphuric acid by mistake. The reason was given that the labels were of exactly the same design and colour, differing only in the name of the acid. While agreeing that the laboratory assistant should have taken more care to read the names, if the labels had each been of differing designs the mistake would probably not have occurred. It was agreed to put an additional distinguishing label on all winchesters of concentrated sulphuric acid to prevent a possible future incident.

Care should be taken not to store together chemicals which are incompatible. A list of such chemicals is given on pages 61–63.

Certain chemicals, such as winchesters of ammonia liquor, can be hazardous on their own. Under warm conditions excessive pressure can be generated and it has been known for the bottles to burst. The winchesters should be kept in cool conditions and opened very carefully, the operator taking the necessary precautions and wearing protective clothing on the assumption that the liquid will spurt out of the bottle. Chemicals should also be issued on a first-in first-out basis as ageing of chemicals can lead to problems. An outstanding example of this occurs in the case of the alkali metals particularly potassium, which on prolonged storage can form a coating of yellow potassium superoxide under which is a layer of potassium monoxide in contact with the metal. This can explode violently on cutting or handling and has given rise to several serious accidents. If the potassium metal has an orange or yellow coating it should be destroyed, preferably by burning in an open coke fire. Alternatively, the metal may be cut into small pieces under xylene and disposed of by the addition of tert-butanol.

Chemicals no longer needed should be discarded. Safe disposal of hazardous chemicals can be a problem and reference should be made to the literature available or the manufacturer of the chemical for the correct procedure. There are several useful reference books giving methods of disposal.[1,2] Thoughtless transporting or carrying of chemicals is frequently observed. Flammable liquids should be carried in flameproof cans, certainly not in the buckets sometimes seen. Trolleys should be used when quantities of chemicals are to be moved, and corrosive chemicals in glass should be carried in proper containers such as winchester carriers. Gas cylinder storage is dealt with in Chapter 7, and a gas cylinder store is shown in Fig. 11.1.

Liquified gases are stored in special containers, and a typical installation for liquid nitrogen is shown in Chapter 1 (Fig. 1.5). When small containers are filled from such a vessel gloves and goggles must be worn, since liquid nitrogen in contact with the skin can produce painful burns. Although non-toxic it can evaporate in a closed atmosphere and in doing so decrease the oxygen concentration to a dangerously low level, and it should therefore be used only in well-ventilated areas. Similar precautions apply to solid carbon dioxide. Hazards with other liquified gases are associated with the particular chemical. Examples include flammability in the case of liquified petroleum gas, toxicity in the case of chlorine and sulphur dioxide, and corrosivity in the case of ammonia, which can also form explosive mixtures with air in the concentration range 15–28 per cent v/v. Those who work in stores should be aware of, and trained in, the correct handling of all the chemicals stored there.

Fig. 11.1. A gas cylinder store. Acetylene gas cylinders are stored in a similar but separate store.

The storage of radioactive materials should be under the control of the Radiological Protection Officer although there is no reason why they should not be located in the laboratory store. They should be kept in their original package in a metal lockable safe labelled with the radiation trefoil. The appropriate fire authority must be informed of the presence and location of radioactive materials. Poisonous chemicals should be in a locked cupboard as mentioned in Chapter 1.

Flammable Solvents and Stores

The storage of flammable solvents, defined in the UK as materials with a flashpoint of less than 32°C and capable of supporting combustion, is covered by statutory requirements in practically every country. Apart from the legal requirements, the insurance carriers will make inspections and may ask for additional precautions. In the UK the local authority in the area concerned is the enforcing body in so far as regulations are involved. They are concerned with the initial approval of plans for a flammable liquids store and the issue of licences where applicable. There are two standards of construction, one applying to petroleum stores and one for highly flammable liquids. The latter may be stored in petroleum stores but petroleum must not be kept in stores constructed for liquids of higher flashpoints. Typical requirements for flammable liquids stores are that they should be separated from main buildings by not less than 4 m in the open air. The floor should have a bund or sill at the entrance door so that the contents of the largest vessel stored plus 10 per cent can be contained in case of leakage. There must be a ramp so that access by trolley is possible. High and low-level ventilation should be adequate to prevent the build-up of dangerous concentrations of vapour. Ventilation apertures should be covered with flameproof gauze and must vent to a safe place. Any source of heating must be guarded so that solvent containers cannot be stacked against it, and must not provide a means of ignition. Construction must be of fireproof materials—4½ in brick is an example—and the roof of light construction. Light fittings and switches must be of an approved flameproof design, fire-fighting equipment must be provided immediately adjacent to the outside of the building and "no smoking—highly flammable liquids" signs in lettering large enough to be clearly seen displayed. Flammable materials must not be transferred from one container to another inside the store and these operations must be done elsewhere. Organic peroxides, cellulose nitrate or other highly toxic non-flammable substances must not be placed in the flammable store.

Empty containers that have contained highly flammable liquid and have not been completely freed of vapour should be kept in the store. No naked flames should be allowed inside the store and if hot work is to be carried out there then HM Inspector of Factories should be notified beforehand so that he can advise on the precautions necessary to comply with legal requirements for safe conduct of the work. A typical flammable liquid store is shown in Fig. 11.2.

Bulk Storage of Flammable Liquids in Tanks

It is preferable to store large amounts of flammable liquid in tanks if this is practicable. The tanks should not be inside or on the roof of a building, but preferably above ground in the open air. Where space is restricted they may be buried below ground but not beneath process buildings.

Fig. 11.2. A store for waste flammable liquids. Note the ramp for easy access, the air vents, the external fire extinguishers and the warning notice on the inside of the door, as the notice on the outside of the door cannot be seen when the door is open. The floor inside is at a lower level to retain any spilled liquid.

The location of tanks above ground in the open is subject to regulations covering distance from adjacent buildings, method of discharge of contents, provision of bund wall, treatment of the ground on which the tank is built, provision of drainage, layout of tanks, and methods of support.

For underground tanks, the regulations cover distance from building line, provision of manholes, siting of filling and dipping connections and vent lines, securing and protecting the tanks from surface loadings, and supporting the tank.

The design and construction of tanks, tank fittings, valves, and pumps are all subject to detailed regulations and suitable level measurement devices must be fitted to each tank.

For full details of requirements regarding the use of tanks for storing flammable liquids the Guidance Note of the Storage of Highly Flammable Liquids, Chemical Safety 2, issued by the Health and Safety Executive, should be consulted.

Apparatus and Equipment

Storage of these items is relatively straightforward and should present few hazards. As far as possible heavy and bulky items should be stored at low levels. Smaller items can be stored at higher levels, preferably in fixed, open-topped containers which cannot fall. Such containers should be clearly labelled so that they can easily be identified. Storage of apparatus in its original packaging prevents damage and facilitates handling.

General

It is most important that a high standard of good housekeeping be maintained in a laboratory store. A methodical system of working is essential: chemicals, apparatus, and equipment must be stored in their correct place when they arrive; packing cases and material should be disposed of as soon as possible to eliminate potential sources of fire. The stores area should be made a no smoking area; this is essential in the chemical store. A drench shower or copious supply of water and sodium carbonate should be available adjacent to the corrosive chemicals; adequate fire-fighting equipment should also be ready to hand. Attention is drawn to the method of storage in which movable racks are used (Fig. 11.3). In small units there is no real danger of non-handicapped persons being trapped between two racks as they are moved by another person, but deaf persons may present problems. This can be prevented by either using a simple flag indicator to show when a person is between two racks or for a locking rod to be inserted between the racks to hold them apart while persons are working between them.

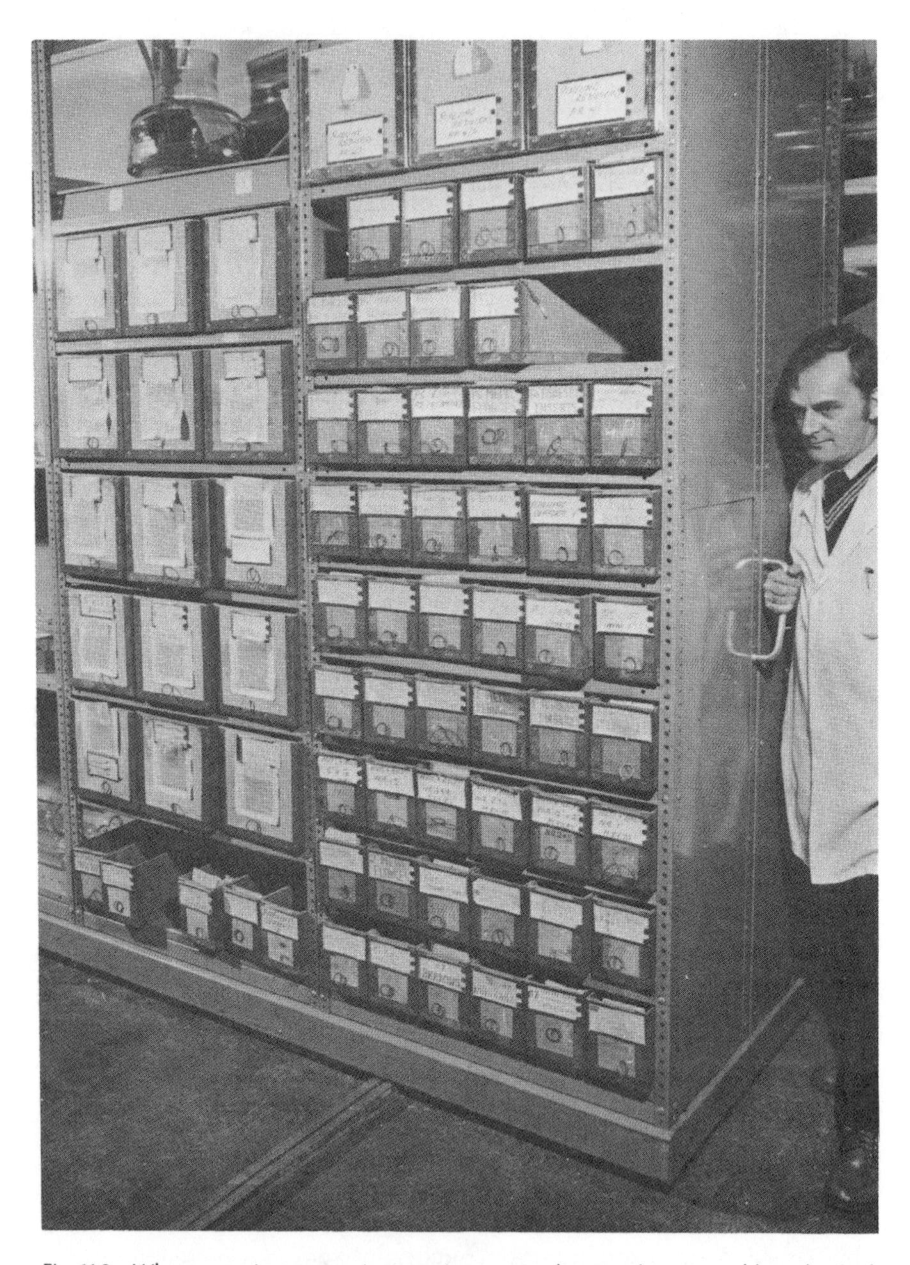

Fig. 11.3. When space is at a premium, equipment can be stored on moveable racks. Such a system is shown here in which laboratory apparatus is stored on the racks which move on rails along the floor. The racks are moved easily and care has to be taken to ensure that persons cannot get trapped between the racks.

ANCILLARY EQUIPMENT

General

The modern laboratory block with its sophisticated equipment requires a very high standard of service facilities. Constant-temperature laboratories require refrigeration as well as heating; clean-air laboratories require complicated electrostatic and filtration systems; other services which may be required are air compressors, distilled or demineralized water plants, towns gas compressors, water pumps, and vacuum engines. Personal services include telephone and paging systems, and canteens. It is usually in these areas rather than in the actual laboratories that the lost-time accidents occur which spoil the safety record of large laboratory complexes. It is in these areas that a higher degree of supervision is necessary to ensure that safe working conditions are rigorously followed. For any laboratory complex of reasonable size it is most advisable to place these service facilities under the control of a qualified engineer who has had experience of running a laboratory block or of working on chemical plants. By this means he will be knowledgeable about statutory or other procedures which must be followed, the safe maintenance of machinery, and so on. Although factory legislation does not normally apply to laboratories and ancillary areas it is wise to operate as if it did. For example, because electrical equipment is in a laboratory it does not mean that work can be carried out on it without a proper isolation and permit-to-work system being used. Experience has shown that, where the suggestions made in this chapter are carried out, a satisfactory safety record can be achieved in offices, laboratory stores, and other buildings.

CANTEENS

Employee eating areas of all kinds are included in this description. Exceptional standards of hygiene must accompany good safety practices. Falls are responsible for the highest number of accidents, and most of them occur when catering workers are walking on the level. A combination of a good anti-slip finish on floors, sensible cleaning routines which get rid of grease and do not leave floors wet for a long period, freedom from obstructions, and the wearing of suitable footwear will all contribute to the lessening of this hazard. A surprising tonnage of goods is handled in quite a short time, which indicates that training in kinetic handling methods is desirable. Where gas is used to fuel cooking equipment there should be a routine check of all gas taps before the kitchen is vacated at the end of a work period. Responsibility for this check should be firmly placed on a named individual. Doors should be left open so that a leaking gas cock cannot form an explosive mixture inside equipment whilst it is not in use.

Cuts, burns, and scalds also occur. They can be controlled by ensuring that knives are not left on work benches, that heat-resistant oven cloths are used to handle hot containers, and that the handles of saucepans and other containers are not allowed to protrude over stoves. When machinery is cleaned it should be completely isolated by removing connecting plugs from sockets where possible, or locking off switches on larger machines. Microwave ovens should be used in accordance with instructions drawn up by the radiation protection officer or similar knowledgeable person. Jewellery should be removed before work starts. Suitable protective clothing includes a one-piece overall, preferably without pockets. Handkerchiefs should not be allowed, but suitable tissues made available in dispensers accompanied by a self-closing disposal bin into which they can be placed after use. Hair should be completely enclosed. All personal protective clothing should be clean and soiled garments changed without question as soon as possible. Each day or shift should end with a complete clean-up of the entire cooking and food preparation areas, and all the equipment that has been used, including drying cloths. Supervisors must check that this is carried out with complete thoroughness and not accept anything but the highest standards. Temperature control is difficult but can be achieved by proper extraction ventilation, with the addition in difficult areas of refrigerated air directed to problem spots. Discussion and mutual co-operation will encourage the development of pride in the standards of safety and hygiene achieved.

REFERENCES

1. Grundy, J. W. and Rosenthal, S. G. (1980). "VDU's and You, Information for Operators", The Association of Optical Practitioners, Bridge House, 233–234 Blackfriars Road, London SE1 8NW.
2. Gaston, P. J. (1970). "The Care, Handling and Disposal of Dangerous Chemicals", Institute of Science Technology, Northern Publishers (Aberdeen) Ltd., Aberdeen.
3. Laboratory Waste Disposal Manual, 1969, Chemical Industries Association, Washington DC.

FURTHER READING

"Care in the Office" (1977). The Royal Society for the Prevention of Accidents, Cannon House, The Priory Queensway, Birmingham.

"Code of Practice for the Keeping of Liquefied Petroleum Gas in Cylinders and Similar Containers" (1973). HMSO, London.

"Code of Practice for the Recovery of Flammable Solvents" (1977). Chemical Recovery Association, Petrol House, Hepscott Road, London E9 5HD.

"Dangerous Substances, Guidance on Dealing with Fires and Spillages" (1972). HMSO, London.

"Fire-fighting Equipment, Fire Alarms and Fire Drills in Offices and Shops" (1963). SHW5, HMSO, London.

"Flash Points" (1962). The British Drug Houses Ltd., Poole, UK.

Grundy, J. W. and Rosenthal, S. G. (1980). "Vision and VDU's", ibid.

Human factors aspects of visual display unit operation (1980). Research Paper 10, HMSO, London.

"Industrial Solvents, Flammable Liquids and Low Melting Point Solids" (1965). Fire Protection Association, Aldermary House, Queen Street, London EC4.

Manos, J. (1980). Stressless Use of VDU's, *Health & Safety at Work* **34,** August.

"Model Code of Principles of Construction and Licensing Conditions (Part II) for Distributing Depots and Major Installations" (1968). HMSO, London.

"Offices, Shops and Railway Premises Act, 1963", HMSO, London.

"The Petroleum (Consolidation) Act 1928 (Enforcement) Regulations. 1979", HMSO, London.

Porter, W. E., Ketchen, E. F. *et al.* (1978). "Health Considerations Relative to the Use of Solvents in the Chemistry Laboratory", *Energy Res. Abstr.* **3,** Part 20, Abstract No. 48907.

"Raw Materials Safety Data Handbook" (1976). Paint Makers' Association of Great Britain Ltd., Prudential House, Wellesley Road, Croydon CR9 2ET.

"The Storage of Highly Flammable Liquids", Chemical Safety, 2 January 1977, Health & Safety Executive, HMSO, London.

12

First Aid

FACILITIES

Every laboratory, or group of laboratories, should be equipped with adequate facilities for administering first aid should injury or illness occur. This need is sometimes neglected in very small locations. A serious examination of these arrangements should be undertaken to ensure that:

a) they are adequate in relation to the degree of risk;
b) proper training has been carried out so that rapid treatment is available when needed and the procedure is known;
c) that the statutory requirements of the country concerned are complied with.

Such requirements exist in almost every country and in many cases include a clause calling for the availability of suitable transport to the nearest hospital. The German Federal Republic laws lay down that all but trivial injuries are treated by a doctor "and not by a quack". In many cases the kind and quantities of material to be on the premises are prescribed. Where this is so it is common for suppliers to sell accurately stocked cabinets to comply with the law.

First-aid Boxes

The siting of first-aid boxes needs careful thought. In a 1000-bed hospital no first-aid boxes were provided as the authorities considered it unrealistic to do so because highly skilled treatment was always available from consultants and residents or interns. Outside specialists carrying out a safety audit pointed out that the laboratories were more than 400 yards from the nearest point at which treatment was available and that a good deal could go wrong while injured workers were transported that distance. First-aid facilities were then installed in the laboratories, laundries and other remote locations on this sprawling site.

There may be risks peculiar to specific laboratories so that the statutory contents of first-aid boxes need to be supplemented. An obvious example is where cyanide is used. Where this chemical or any of its compounds is present then a suitable antidote must be rapidly available. One example noted was in a laboratory in which work on potassium cyanide and similar compounds was occasionally carried out. The antidote was kept in the surgery of the village doctor almost a mile away. This was because it was felt only a doctor should administer it. The grave mistakes here were not only the time lag between absorption from one of the many points of entry into the body (respiratory, gastrointestinal, unbroken skin) but the possibility of the doctor being absent from his surgery when needed. Another frequent omission is an instruction book. A fire occurred in a small offices/laboratory complex. Two workers were affected by smoke inhalation and burns and, although a well-equipped first-aid box was accessible, no-one knew what to do. The result was that the injured persons' condition worsened before firemen arrived and rendered first aid. An instruction book, though not a replacement for training, would have enabled something to be done.

Other Equipment

The first-aid box or cabinet should be supplemented by a stretcher, blankets, splints, and a haversack. The last mentioned should contain emergency equipment such as scissors, tweezers, tongue depressor, dressings, cottonwool, and a simple hand-operated resuscitator. Where an injured person is immobilized this haversack can be taken to the patient so that necessary first-aid is given before movement is begun. A first-aid room, medical centre or treatment room, dependent on the number employed and degree of risk, may be needed on a statutory or commonsense basis. For instance, laboratories with easy and rapid access to a well-equipped hospital casualty department will need a less elaborately equipped room than others in a relatively isolated position or with hazards of a high order. In any case certain basic rules apply. The floor and walls should have a smooth impervious finish. Lighting should be good and adjustable, and hot and cold water available, coupled with adequate heating and ventilation. The location should be at ground-floor level, sanitary facilities provided, and at least one doorway should be of ample width to take a stretcher. A cupboard should contain dressings, antidotes, a sterilizer, and there should also be an examination couch and chair, and a waste bin. Respiratory equipment and resuscitators will be provided in relation to the risk. Personal protective respiratory equipment for first-aiders is often forgotten. They may have to take the first-aid haversack into laboratories where toxic fume is, or has been, present. Injured workers may have clothing saturated with fumes. The help first-aiders can give is sometimes restricted because of their own lack of personal protection.

Transport of Injured Employees

It may be necessary to transport an injured worker to hospital. Very large locations may have an on-site ambulance, others a vehicle such as a personnel carrier which can be adapted for use as an ambulance. Remote locations will clearly have a greater need for suitable transport than will those near to hospitals and ambulance stations. In any case, emergency telephone numbers clearly displayed in the first-aid room should include those of the nearest doctor and hospital. Ambulances will normally be available through the police emergency call number. Patients sent to hospital should have a label affixed to their clothing stating the nature of the incident, toxic or corrosive materials (if any) involved, and details of the treatment given so that the correct follow-up action can be taken at the hospital without loss of time. Purpose-made labels of this kind are obtainable in some countries, e.g. from the Chemical Industries Association in the UK. Casualty department doctors can be involved in discussions before incidents occur so that they are made aware of the kind of emergency they may have to deal with. It is unrealistic to expect casualty officers to be instantly knowledgeable about treatments for obscure contacts or ingestions, and preliminary discussions may save valuable time in an emergency. A chemist inhaled dust in considerable quantity when a large container of particulate matter collected via the extraction system in a chemical works was sent to the laboratories for analysis and was accidentally overturned. Vanadium pentoxide was used as a catalyst in the plant concerned and the chemist experienced considerable distress as convulsive coughing was induced. When the company concerned first used vanadium pentoxide they commissioned a report from a professor in the faculty of medicine at a local university. The report included details of treatment following inhalation. A copy of this part of the report was sent to hospital with the injured man and appropriate treatment was started immediately. The duty casualty officer said afterwards that he had no knowledge of the correct procedure and, but for the information supplied, would have needed to seek help from colleagues and literature available, while the injured man was in considerable distress. The employer concerned subsequently supplemented the labelling system with discussions as suggested earlier in this chapter.

Eyewash Stations and Emergency Showers

Adjuncts to first-aid services are eyewash stations and emergency showers.

In small, low-risk laboratories the former may consist of eyewash bottles suitably placed away from other substances so that the wrong bottle cannot be picked up by someone whose sight is temporarily impaired. Cleaning, testing, and topping up should be a routine matter. In larger and/or more hazardous locations eyewash fountains are desirable. The most effective treatment for harmful chemicals in the eyes is rapid, thorough, and prolonged washing using copious amounts of water. This is best achieved by fountains which direct a gentle stream of water on to the face and around the eyes. Water should not be poured directly on to the eyeball, but on to the surrounding area, preferably the bridge of the nose, from where it will run on to the eyeball. A problem with sufferers from eye injuries is that frequently they are in a state of panic and may have to be forcibly held over a fountain. When this is not possible any conveniently available utensils can be used—a cup, a beaker, or any clean container which Till hold water—and the water poured on to the area around the eyes. A common fault in applying first-aid of this kind is stopping too soon. Irrigation of the eyes should continue for several minutes. Deluge showers are also needed. On small sites a possible, though not entirely satisfactory, substitute is the normal shower cubicle provided for workers' ablutions. If it is used for washing corrosives etc. from the skin the cubicle must be thoroughly cleaned before reverting to normal use. Purpose-installed deluges can be placed strategically in corridors with the floor gently inclined to a drain immediately beneath the shower. Portable shower cabinets, with a self-contained water supply ejected by air pressure when a worker steps on a platform have their uses. Operation by stepping on a plate which automatically opens the water valve is preferred to hand operation where possible. It is not likely that the problems of frozen pipes will arise in laboratories but consideration must be given to the possibility.

Training

Various training manuals can be recommended for those who wish to acquire an elementary knowledge of first-aid through self-study. The same manuals are also excellent when formal training is undertaken. The British Red Cross, the St. John Ambulance Brigade manuals published by the organizations mentioned, and "First Aid in the Factory" published by Longmans, Green and Company, London are recommended in the UK. The United States Bureau of Mines Manual of first-aid instruction and the American Red Cross first-aid textbooks are useful in the US. Wherever possible members of staff should undertake formal training so adequate cover is always available. There is normally a readiness to do this on a voluntary basis though some employers choose to recognize trained first-aiders by a token annual payment. A worthwhile idea operated by at least one large chemical company is to incorporate first-aid training in the general programme relating to trainees entering laboratories from school or college. Whatever kind of training that may be, the name(s) of the workers concerned should be clearly displayed on every first-aid cabinet.

It may be thought that some of the suggestions made are unrealistic in so far as the small laboratory is concerned. Those in charge of laboratories, whether they be a major complex or a one or two-person operation, should look closely at first-aid arrangements and pose the questions set down at the beginning of this chapter. Until a satisfactory response to this self-examination is forthcoming, then changes should be made until a situation is reached in which it can fairly be stated, should an injury occur, that everything reasonable has been done through preplanning to ensure prompt and correct first-aid, and other measures to minimize the effect.

ACCIDENT PROCEDURE

THE FIRST DUTY OF ANY PERSON CONNECTED WITH AN EMERGENCY IS THE PRESERVATION OF HUMAN LIFE.

Minor Injuries

Every accident or injury, however slight it may appear to be, must be treated in the First-aid Room. The neglect of even a small scratch may have serious results.

Major Injuries

In the case of a serious accident, first take obvious action, such as extinguishing fires, dealing with corrosive materials and, if there is personal injury, arranging immediate medical aid. Do not move an injured person except to a position of less danger. Cover him to keep him warm and so lessen the effect of shock. Ensure that full details of an accident or a near-accident are given to the safety officer so that an investigation can begin as soon as practicable.

Electric Shock

Switch off the current before attempting to rescue a person in contact with a live circuit. If this is not possible, protect the hands with rubber gloves, or dry woollen material. Before touching the person stand on a dry mat or coat to increase the insulation. Send for qualified medical aid immediately such an accident occurs, as artificial respiration and treatment for burns and shock may be necessary.

The smallest current which can be detected through the skin is generally considered to be approximately 1 mA r.m.s. at 50 Hz a.c. and 5 mA d.c. (the tongue is considerably more sensitive). On increasing the current a stage is reached when severe muscular contractions make it difficult for the casualty to release his hold. This stage occurs with currents of about 15 mA at 50 Hz a.c. and 70 mA d.c. The effect increases with frequency, e.g. only 7 mA are required at 60 Hz but this does not extend to very high frequencies.

Increase in current to beyond 20 mA 50 Hz or 80 mA d.c. brings danger to life. At 100 mA for both a.c. and d.c. irregular contractions of the heart occur which are almost certain to be fatal. Currents of 1 A and above through the body cause severe burning.

The resistance of the body varies enormously according to the conditions and from person to person. It can be as high as 10000 ohms or as low as a few hundred. Even with a resistance of 10000 ohms a 240 V a.c. supply will result in a current of more than 20 mA which could be lethal. In fact, death has been recorded from only 60 V.

The above remarks apply to current passing through the body, e.g. from hand to hand or hand to foot. It is possible for part of the body, e.g. a finger, to short-circuit a supply. This will not result in an electric shock but will cause a severe burn requiring medical treatment.

Poisons and Poisonous Gases

In any case of poisoning, speed of action is essential: SEND FOR MEDICAL AID IMMEDIATELY AND KEEP THE PATIENT WARM AND QUIET.

Where it is clear that medical aid will be delayed, treatment may be administered according to the list below. Where the action of the poisonous substance is very rapid, e.g. with cyanides, the antidote must be available when such substances are being used. Never give anything by mouth to an unconscious person.

Cause of poisoning unknown. There are three general principles which can be followed.

1. Give large quantities of warm water.
2. Give activated charcoal if ingestion is possible.
3. With gassing casualties remove patient to fresh air, keeping him lying down and warm, and loosen clothing at waist and neck. Give oxygen if necessary but only employ artificial respiration if breathing actually stops.

POISON	TREATMENT
Acids:	
Acetic Hydrochloric Nitric Perchloric Phosphoric Sulphuric Thioglycollic	Wash mouth thoroughly with plenty of water or 5 per cent sodium bicarbonate. Give milk of magnesia and plenty of water or milk to drink.
Oxalic (also oxalates)	Give immediately a suspension of chalk or magnesium in water or a solution of calcium lactate.
Hydrocyanic or Prussic (also Cyanides)	If inhaled give amyl nitrite to inhale for 15–30 s. Repeat every 2–3 min. If swallowed give cyanide antidote immediately if patient is conscious. *Antidote* The following solutions must be ready for use wherever cyanides or prussic acid are used. A. 158 g BP ferrous sulphate ($FeSO_4\ 7H_2O$) and 3 g BP citric acid dissolved in 1 litre of cold distilled water. (Discard when brown discoloration occurs.) B. 60 g sodium carbonate (anhydrous) dissolved in a litre of distilled water. For use, mix about 50 ml of A and B and swallow the mixture. In either case apply artificial respiration if breathing has stopped. Where there is a risk from these substances a supply of dicobalt edetate should be available for intravenous injection, which can only be done by a medical officer.
Hydrofluoric	Give weak lime water and then warm water. Repeat three or four times. Give artificial respiration if necessary. With burns on skin, drench with water, remove contaminated clothing and dip affected parts in, or swab down with a solution of sodium bicarbonate.
Alkalis:	
Ammonium hydroxide Potassium hydroxide Sodium hydroxide Sodium peroxide	Give very dilute acetic acid or fruit juice and then milk.
Antimony salts	Give tannic acid or strong tea.
Arsenicals	Give Epsom salts, then egg white and/or milk.
Barium salts	Give Epsom salts.

Bromine	If taken orally rinse mouth with 3 per cent sodium carbonate and magnesia in water. Then give milk and a suspension of 10 g of magnesium in 150 ml of water. If inhaled give absolute rest. Oxygen may be given but do *not* use artificial respiration. With burns on skin, drench with water and bathe with a dilute solution of ammonia or sodium thiosulphate.
Carbon disulphide	Artificial respiration and oxygen.
Carbon tetrachloride	Give Epsom salts. Do *not* give oils or fats. Artificial respiration if necessary.
Carbon monoxide Acetylene Coal gas	Remove patient to fresh air, keeping him warm, and employ artificial respiration. Give oxygen if respiration slow or irregular. Do *not* walk the patient about or give stimulant.
Chlorine	Remove patient to fresh air, keeping him warm and give absolute rest. Administer oxygen if breathing is weak. If taken orally as chlorine water, rinse the mouth with 3 per cent sodium carbonate and magnesia in water. Then give milk and a suspension of 10 g of magnesia in 150 ml of water.
Chloroform Ether	Artificial respiration and oxygen if necessary.
Hydrogen peroxide	Give Epsom salts.
Hydrogen sulphide Hydrogen selenide (The effects of weak concentrations are cumulative)	Give artificial respiration and persist, if necessary, for several hours. Oxygen may be given.
Lead salts	Give Epsom salts. Give white of egg or milk and a stimulant.
Mercury salts	Give large quantities of water. Give milk and a stimulant.
Nitrous fumes (may show delayed action)	Give absolute rest. Administer oxygen if necessary.
Phosgene (may show delayed action)	Give absolute rest. Administer oxygen if necessary. Do *not* use artificial respiration.
Potassium permanganate	Give pure charcoal, egg white, milk and a stimulant.
Selenium and vanadium compounds	For inhalation of dusts and vapour give absolute rest and keep warm. If swallowed, wash out mouth thoroughly with water.
Silver salts	Large amounts of water. Then give egg white or milk.

FURTHER READING

Hooper, E. (1980). "The Safe Use of Electricity", Royal Society for the Prevention of Accidents, The Priory Queensway, Birmingham, UK.

Deichmann, W. B. and Gerarde, H. W. (1972). "Toxicology of Drugs and Chemicals", Academic Press, London and New York.

Dreisbach, R. H. (1971). "Handbook of Poisoning; Diagnosis and Treatment", Lange Medical Publications, Los Altos, C.A., USA.

Eagers, R. Y. (1969). "Toxic Properties of Inorganic Fluorine Compounds", Elsevier Publishing Co. Ltd, London.

Effects of Current Passing through the Human Body, International Electro-

Technical Commission, IEC Report, Publication 479, 1st Edition, 1979, Bureau Central de la Commission Electrotechnique International, Geneva.

First Aid in Factories. Health, Safety and Welfare Series No. 36, HMSO, London.

Guy, K. (1965). "Laboratory First Aid", Macmillan, London.

Harvey, B. and Murray, R. (1958). "Industrial Health Technology", Butterworths, London.

Hunter, D. (1978). "The Diseases of Occupations", Hodder & Stoughton, London.

"Industrial Dermatitis", Chemical Industries Association, London.

Laboratory First Aid Wall Chart, and Spillages of Hazardous Chemicals Chart, BDH Chemicals Ltd, Poole, UK.

Matthew, H. and Lawson, A. A. H. (1979). "Treatment of Common Acute Poisonings", Churchill Livingstone, London.

Paltry, F. A. (1963). "Industrial Hygiene and Toxicology", Interscience, New York and London.

Porter, W. E., Ketchen, E. F. *et al.* (1978). Health considerations relative to the use of solvents in the chemistry laboratory, *Energy Research Abstracts* **3,** Part 20, Abstract No. 48907.

Trevethick, R. S. (1976). "Environmental and Industrial Health Hazards, a Practical Guide", William Heinemann Medical Books, London.

"Safety Precautions in the Use of Electrical Equipment", Imperial College of Science and Technology, London.

13

Legal Aspects in the United Kingdom

"I know not whether laws be right Or whether laws be wrong" wrote Oscar Wilde in his ballad of Reading Gaol. Until recent years, people who worked in research and development laboratories were not unduly concerned about filling this gap in their knowledge had it existed. Few laws applied to them. The preoccupation with the terms of the Factories Act which consumed some of the working hours of a factory manager was not a burden to the research and development manager. The complexities of the Highly Flammable Liquids and Liquefied Petroleum Gases Regulations (HFL and LPG Regs.) caused no loss of sleep to him, because the work areas he controlled were exempt from their requirements.[1]

Much industrial legislation applied only to activities which were directly related to production. Thus quality control laboratories with their intimate link with day-to-day factory operations were considered to be subject to the same legislative control as was the factory. The research and development activities were outside a great deal of the law intended to control the way in which peopled worked, their environment, their health and safety, although it would be difficult to envisage a successful production unit not backed up by adequate R and D.

This anomolous situation was perpetuated until The Offices, Shops, and Railway Premises Act 1963 (OSRA) became operational. The term "office premises" is defined as any building, or part of a building, used solely or principally for office purposes. So offices in laboratories, irrespective of the kind of activity carried on therein, were brought within the scope of this particular Act. This only partly put right a situation not only perplexing but also stupid. For example, prior to 1963 a cleaner was working in an office block half the length of which was occupied by works personnel, the remaining half by R and D people. She was using a small but perfectly adequate working platform with two steps up to the standing surface to reach the fanlights she was cleaning above each office door. Having cleaned the last fanlight in the works section she moved her platform 1.5 m sideways to clean the first fanlight in the R and D section. For some unaccountable reason she fell while doing so and broke her left leg. The accident was not reportable, no one was interested in investigation except internally by the occupier, no laws were broken. Considering that had the incident occurred some 1.5 m away and a few seconds earlier all kinds of official activity would have been stimulated, the situation can only be described as simply daft. The OSRA put an end to such absurdities (the incident became reportable on Form OSR 2 under its terms) but only in a fairly limited field. The mainstream of many laboratory operations continued without statutory constraints until The Health and Safety at Work etc. Act 1974 became operational. The application of this Act to all persons at work except domestic servants in private employment (Section 51) brought within its scope an estimated 8 million workpeople not previously covered by legislation, including laboratory workers.

HEALTH AND SAFETY AT WORK ETC. ACT 1974

The Health and Safety at Work etc. Act (HASAWA) therefore becomes the linchpin which holds together the legal framework within which laboratories must be operated. Before the main provisions of this Act are briefly set down the position with regard to schools, hospitals, Crown property, and employees working in these places must be outlined. All the requirements of HASAWA Part I apply to these locations and persons except sections 21–25, which deal with improvement and prohibition notices etc., and sections 33–42, which are related to offences. The intention is that employees of the Crown should be self-regulating; the sanctions which apply to others should apply to them also but by internal rather than external processes. So an inspector will not issue an improvement notice concerning an educational establishment, for instance. He will draw the attention of the authorities to the circumstances or conditions which would give rise to an improvement notice in other establishments, and ask that the necessary remedial action be taken. It is then a matter for internal action to see that this is done. The same considerations apply to prosecutions.

In the early days of HASAWA there was a lack of clarity as to exactly how the legislation applied to hospitals and schools. One trade union with a substantial membership in hospitals took counsel's[2] opinion in 1975, with regard to the application of section 33–42 to hospitals. The opinion was that they did apply and employees in such places were liable to prosecution if any offence specified in the Act was committed by them. In the event no legal proceedings have been taken and the internal action outlined has presumably been followed. The Health and Safety Commission, the body responsible in broad terms for the administration of HASAWA is not happy with this situation, a view shared by others. It is a possibility that amending legislation will in the future extend all or most of the sections mentioned so they apply to the Crown in the same way as to others.

The requirements of HASAWA are set down in an abbreviated form for ease of understanding. What follows is not to be taken as a legal interpretation but as a guide, a qualification which applies to all the summaries in this chapter.

It is the duty of every employer to ensure as far as is reasonably practicable, the health, safety, and welfare at work of all his employees. (2(1)). Failure to comply with this general duty—a matter of opinion many consider—has been responsible for a high proportion of all prosecutions under the Act. Frequently it is thrown in as a kind of makeweight in addition to an alleged breach of a specific duty. This duty, although expressed in general terms, overrides the specific duties which follow. An employer must:

1. Conduct his undertaking in such a way as to ensure that persons not in his employment who may be affected thereby are not exposed to risks to their health and safety. (3(1)) is particularly relevant as far as laboratories are concerned. See our later comments regarding chlorine release.
2. Provide and maintain plant and systems of work that are safe and without risks to health. (2(2)(a)). It is not too difficult in a well-organized laboratory to ensure that plant and equipment is up to standard. Systems of work are not always good. Entrenched ideas can perpetuate the use of outdated systems which need revision. Permit-to-work systems should also be used more extensively than hitherto.
3. Make arrangements for safety and absence of risk to health in connection with the use, handling, storage and transport of articles and substances. (2(2)(b)). Carrying highly flammable liquids around in open-top vessels (even buckets) has been often seen. Thoughtless storage in unmarked and unsafe containers will not do either!
4. Provide such instruction, information, training, and supervision to ensure the health and safety of employees at work. (2(2)(c)). Communication is not something at which many managers excel. It is a waste of time to instruct, inform, and train people who at the end of the day are no wiser. Those who cannot communicate well can improve if they take the trouble to learn the necessary skills.
5. Maintain the place of work and means of access and egress in a condition that is safe. (2(2)(d)). This does not mean standing on chairs or stools to get to out-of-reach materials, or cluttering up floors or benches with equipment or spillages.
6. Provide a working environment that is safe, without risk to health and with adequate welfare facilities. (2(2)(e))

It is not necessary for an inspector of an enforcing authority to show that an employer is in breach of any of the specific duties set down if there are indications that there is a breach of Section 2(1). Where welfare is mentioned in this or any other legislation the reference is only to welfare facilities required by law. An employer might sponsor leisure-time activities for employees, or special help for those who are sick. Such commendable activities are not "welfare" in the legal sense.

REASONABLE PRACTICABILITY

Many sections of HASAWA, including all the above-mentioned, are qualified by the phrase "so far as is reasonably practicable". Its meaning is clear in some cases, obscure in others. An example of clarity is in Regulation 3(4) of the Safety Representatives and Safety Committees Regulations (dealt with later) which requires that safety representatives will have been in the employment concerned, or similar employment, for two years so far as is reasonably practicable. It does not require an Einstein to conclude that if no employees exist with this service qualification then someone else can be appointed. Other examples are unhappily incapable of an easy interpretation. The generally accepted definition is that pronounced by Lord Justice Asquith[3] who said:

> Reasonably practicable is a narrower term than physically possible and seems to me to imply that a computation must be made . . . in which the quantum of risk is placed on one scale and the sacrifice involved in the measures necessary for averting the risk (whether in money, time, or trouble) is placed in the other. If it be shown that there is a gross disproportion between them—the risk being insignificant in relation to the sacrifice—the defendants discharge the onus upon them.

Many managers feel that even this does not help them too far along the road to complete comprehension, but it is the best we have. A useful rule of thumb is "if in doubt decide on action rather than inaction". A final comment on this somewhat contentious matter: section 40 of HASAWA provides for prosecution where there is failure to comply with a duty or requirement to do something so far as is reasonably practicable. In such a case the onus of proof rests on the accused. It is not necessary for the prosecution to show failure. The defence must satisfy the court that it was not practicable to do more than was done, or there was no better practicable means than was in fact used. There is a tinge of suspicion here that our long-held concepts of innocence until proved guilty and the onus of proof resting on the prosecution are at least part of the way out of the window.

POLICY STATEMENTS, EMPLOYEES' RESPONSIBILITIES

Section 2(3) of HASAWA as amended by SI 1975 No. 1584 requires that where there are five or more employees each employer must prepare a written statement of his general policy with respect to health and safety at work and bring it to the notice of employees. The policy document[4] must be revised as necessary and set down the organization and arrangements for carrying it out. There is a possibility that many small laboratories having the required number of employees, particularly those in the non-industrial section (defined later), are unaware of, or have ignored, this requirement. Eventually the appropriate enforcement authority will catch up with them. Be warned!

Certain obligations are placed on employees. Section 7 calls on every employee to take reasonable care for the health and safety of himself and others, and to co-operate to enable statutory duties to be carried out. There is some misapprehension about this. "Employee" does not mean the laboratory assistant at the bench, the cleaner, or the janitor. It means anyone who works under a contract of employment. With the exclusion of part-timers who work 16 hours per week or less[5] (and some of these may have a contract on a voluntary basis) this means virtually everyone including directors, laboratory managers, and section leaders, some of whom erroneously regard themselves as outside the "employee" definition because their place in the hierarchy indicates that the correct appellation is manager—not so. In fact Section 7 places a much more onerous duty on managers than others. The requirement to take reasonable care for the health and safety of others to non-managers applies to the people with whom they work in its normal day-to-day application. Managers have a wider duty in the sense that "other persons" means all those who come within their area of control. There have in fact been several prosecutions of managers for failing to comply with this section. A requirement that no person shall recklessly or intentionally interfere with anything provided as a result of any statutory provision set down in Section 8 rounds off the employee responsibilities.

It is a rarity but not a complete impossibility to find self-employed persons in laboratories. Sections 3 (2) and (3) have special application to them. They lay down that the self-employed must conduct their business in such a way as to ensure that neither they or any other persons are exposed to risk to their health or safety and that they may be required to give prescribed information to anyone who may be affected by the way they carry on their business. Other relevant sections do, of course, apply to the self-employed.

PERSONS IN CONTROL

There are specific obligations placed on persons in control of premises. Such persons have a duty to ensure that access to and egress from the premises are safe and present no risk to health (Section 4(2)). They are also under a duty to prevent the emission of offensive substances into the atmosphere. The possibility of emissions of this kind is particularly associated with laboratories and special care is needed.

COMPLEXITIES OF SECTION 6

Section 6 has much complexity. Seminars, papers, and guidance notes have all been used to help employers comply. It is the authors' belief that the cause of many of the problems which have arisen is over-elaboration.

To deal first with the detail of the section—manufacturers, designers, importers, and suppliers of articles or substances for use at work are required to ensure that they are safe and without risks to health; to arrange for tests and examinations to ensure this; to carry out necessary research, and to provide adequate information. Important in the context of laboratories is the proviso that the manufacturer etc. is not required to repeat research or examinations carried out by others if it is reasonable for him to rely on their results. Similar obligations are placed on the erectors or installers of any article for use at work.

There is a two-way aspect involved here. Laboratories will be required to undertake a great deal of the investigative and research work called for and to supply detail so that the employer can fulfil his obligations. There is also the need to ensure that incoming information regarding articles and substances used in laboratories is passed on to persons employed. The need for simplicity is paramount. The information requirement is expressed in this way:

> (the employer is) to take such steps as are necessary to secure that there will be available in connection with the use of the article (substance) at work adequate information . . . to ensure that . . . it will be safe and without risks to health.

The necessary steps have not been defined. Clearly to ensure this duty is carried out the information must be presented in such a way that it can easily be understood. Otherwise the adequacy may be questioned. Chemists tend to think that terminology and formulae which are part of the everyday communications within a laboratory are as readily understood in other locations: they are not. Bearing in mind that the reader of Section 6 information is going to be in many cases someone with minimum scientific qualifications it is not very useful to quote a formula, or use scientific terminology.

The same applies, although less frequently, within laboratories where cleaners, security people, and attendants of various kinds are employed. What these people need to know is what an article or substance is, what it can do to them, the precautions they should take and the treatment should they be affected. Think in these terms and discharge Section 6 duties accordingly in everyday language. Details of research and tests carried out can be added although it is the opinion of the writers that such information may be highly technical and therefore perhaps best passed on to managers and safety representatives while still bearing in mind the need for it to be readily available to all. Finally, the protection given to employees by Section 6 applies only when articles or substances are properly used, and it is not necessary to pass on information in such a way that, for example, trade secrets are revealed, so long as the spirit and intention are achieved.

MANAGERS AND CONNIVANCE

Powers possessed by inspectors include the issue of improvement notices requiring a named improvement to be completed within a specified time and prohibition notices which direct that activities named shall not be carried on until matters specified have been put right. There are general duties regarding those who are not employees using the premises in Section 4. Section 37 is of special interest to managers. This states that where an offence under any of the relevant statutory provisions is proved to have been committed with the consent or connivance of, or due to the neglect of, any director, manager, secretary, or similar officer of the body corporate he as well as the body corporate shall be guilty of that offence. There is an acceptable definition of connivance which is "deliberately closing one's eyes to events of which the participant would prefer to remain unaware". The level of management at which this section becomes applicable is not clear. In a much quoted case in Scotland[6] the Director of Roads employed by a district council was convicted following a prosecution under Section 37. There was a good deal of speculation as to whether the man concerned was sufficiently elevated in the management hierarchy to be caught by this section. The conviction was upheld on appeal so the assumption is that he was. Accordingly it is suggested that laboratory managers and also section managers in the larger establishments come within its orbit. Remember that this is a generalization. A good deal depends on the scope and range of responsibility of lower echelon managers.

SAFETY REPRESENTATIVES AND SAFETY COMMITTEES

What has been set down regarding HASAWA is a brief summary of a voluminous and complex piece of legislation; it does not purport to be other than selective. The sections dealt with are particularly relevant and the seeker of further information should study the Act in its entirety. An important set of regulations issued to implement Sections 2 (4) (5) and (6) are the Safety Representatives and Safety Committees Regulations 1978, somewhat unkindly described by some critics as the most loosely worded and imprecise regulations to emerge for many years. Perhaps this is because it was intended that the regulations could be used as a framework within which trades unions and management could work out mutually acceptable implementation arrangements in circumstances in which it would be extremely difficult to legislate in a precise way. Before discussing this a brief summary of the regulations is set down. Increasing unionization among professional and quasi-professional people makes an understanding of them important in laboratories.

Recognized trades unions may appoint safety representatives in workplaces where one or more members of the relevant union(s) are employed. There is no restriction on the number to be appointed and where disagreement on this or any other matter arises it is suggested that it be dealt with through the normal industrial relations channels. Once appointed safety representatives remain in office until they resign, or their appointment is terminated by the union which appointed them, or they cease to be employed in the workplace to which they were appointed. They have functions which include investigation of actual or potential accidents, dangerous occurrences or industrial diseases, and complaints by the members they represent. Their "inspection of the workplace" rights embrace regular inspections not more frequently than three-monthly and additional inspections when major changes occur in the workplace, or new information emerges from the HSE, or a reportable incident occurs. Information of a very comprehensive nature must be made available to them, they have no legal responsibility for acts or omissions in relation to their functions (they must of course observe Sections 7 and 8 of HASAWA), they must be given reasonable time and facilities to carry out their functions, undergo training, and be paid in accordance with a schedule accompanying the regulations. If two or more representatives request a safety committee it must be set up. There is a right of appeal to an industrial tribunal should it be felt that insufficient time is being allowed for the performance of functions or a shortfall in payments, and compensation can be awarded if the appeal is well founded.

This very brief summary of these important regulations is set down because the correct implementation of their requirements can make a useful contribution to improved standards of health and safety in a laboratory. Loose application can lead to contention, time-wasting, and even antagonism. Among the criticism levelled at the regulations is the failure to define "change", "employee complaint",[7] or "investigation on own initiative". Because of alleged loose wording what is known as the "Friday afternoon syndrome" has grown up—a sudden Friday afternoon desire to carry out an additional inspection. In spite of criticism the regulations, properly implemented by co-operation between representatives and management, are a means for progress. Experience has shown that where industrial relations are good this progress has been evident. A final point to be made is that in non-unionized locations there is absolutely no reason why safety committees, joint safety inspections, and so on should not operate on a voluntary basis. There is a whole mass of other legislation in the form of Acts of Parliament and Statutory Instruments, much of which applies to laboratories. An HMSO publication "Health and Safety Executive—Publications Catalogue" contains details as well as other useful information. Several important points are listed below with comments on sections or regulations which experience indicates cause some complexity.

FACTORIES ACT 1961

This applies to laboratories in which work is carried out directly related to the activities which have caused the site to be designated a factory. The definition of a factory appears in Section 175 of the Factories Act. Some analytical, certainly quality control laboratories, are examples. Responsibility for compliance with the Factories Act rests on the occupier of the designated factory, not the employer. The FA requirements impose an overriding responsibility on the occupier who cannot designate this responsibility to a manager or any other person. His responsibility generally speaking extends to all persons on the site whether or not they are his employees. One example of the occupier's overriding responsibility is the provision of safe access and egress irrespective of who employs persons in a particular location. If a contractor has people working in a laboratory subject to the FA on (say) an extension to the building, the responsibility for carrying out the Construction Regulations rests with the contractor. However, the occupier still retains responsibilities which he may have under the FA. The sections most likely to affect laboratory work deal with welfare matters like lighting, heating, washing and toilet facilities, ventilation etc.; passages, stairs and means of access; painting and cleaning; drainage and the disposal of waste; provision of first-aid facilities and first-aiders; guarding, use, and maintenance of machinery; processes producing harmful or explosive dusts; work in confined spaces; and administrative duties.

OFFICES, SHOPS, AND RAILWAY PREMISES ACT 1963 (OSRA)

Laboratory offices on a site designated as a factory, whether separate from a factory or within its curtilage, are subject to this Act in common with all other offices wherever they may be located. OSRA contains a lengthy definition of an office and it can be accepted that any laboratory office comes within it. Responsibility for compliance rests with the occupier except in the case of a multi-occupancy building. Then each occupier is responsible for observance in his own part of the building except for some fire protection arrangements. The owner of the building is responsible for those parts in common use and for some fire protection arrangements. There is a great deal of common ground between the OSRA and the FA.

A SELECTION OF OTHER STATUTORY REQUIREMENTS APPLYING TO LABORATORIES

Fire Precautions Act 1972: Regulations made under this Act effective from 1 January 1977 substitute more stringent requirements for those contained in the FA or OSRA. A fire certificate is required where more than 20 persons are employed, or more than 10 at other than ground level. All places of work are inspected and a fire certificate issued, or refused—until certain work is carried out. The fire certificate sets down certain conditions which must be observed. In the smaller non-certificated locations there are simpler regulations calling for certain fire protection etc. and means of escape.[8]

Abrasive Wheels Regulations 1970 SI 1970 No. 535: Where there is a laboratory workshop and a grinding wheel is used, however small, these regulations apply.

Asbestos Regulations 1969 SI 1969 No. 690: At the time this book was written these regulations were under review. Useful supporting literature can be obtained from the Asbestos Research Council.[9]

Control of Pollution Act 1974: This pulled together a mass of statutory requirements relating to gaseous and liquid effluents, noise, waste disposal, etc.

Chemical Works Regulations 1922 SI 1922 No. 731: Applies to laboratories which are part of a factory defined as a chemical works in accordance with these regulations.

Electricity Regulations 1908 SRO 1908 No. 1312 and Electricity (Factories Act) Regulations 1944 SRO 1944 No. 739: These, too, apply to laboratories which are part of a factory. The "Explanatory Memorandum—Electricity Regulations Form SHW 928" is a useful guide.

First-aid Boxes in Factories 1959 SI 1959 No. 906. First-aid Boxes (Standard of Training) Order 1960 SI 1960 No. 1612: These supplement the first-aid requirements expressed in Section 61 of the FA and Sections 24 and 27 OSRA. Eventually comprehensive first-aid legislation will cover virtually everybody at work including those in educational establishments. Future changes will be of particular importance in connection with laboratories in such places. Even though existing specific regulations may not cover these establishments, failure to comply with their intentions may be construed as a breach of Section 2(1) of HASAWA.

Highly Flammable Liquids and Liquefied Petroleum Gases Regulations 1972 SI 1972 No. 917: Although it has already been pointed out that these regulations do not apply to research laboratories, employers at these places of work ignore them at their peril because of Section 2(1) HASAWA. Review in the 1980s is probable, partly because compliance with these regulations is at an unsatisfactory level. Storage and handling of HFLs in laboratories tends to be particularly poor. There is an excellent guide to these regulations.[10]

Ionizing Radiations (Sealed Sources) Regulations 1969 SI 1969 SI 1969 No. 808. Ionizing Radiations (Unsealed Radioactive/Sources) Regulations 1968 SI 1968 No. 780. Radioactive Substances Act 1960: All have a good deal in common. They apply almost entirely but not exclusively to premises designated as a factory.[11]

Statutory control of laboratory operations is now much more evident since the advent of HASAWA. A comprehensive survey of all legislation has not been attempted, but that which has particular relevance has been identified and attention drawn to sections or regulations which cause particular concern in laboratories. Further guidance is contained in publications recommended in the bibliography.

COMMON LAW DUTY

Statutory duties are set down in Acts of Parliament and Statutory Instruments. Persons having responsibility for discharging these duties can purchase a copy of a relevant legal code and, if they are able to understand it (which is not always easy) know what the law requires of them. Common law duty must also be complied with. Because it is expressed in general terms it is less easy to understand. A reasonable comprehension of this duty is vital. A breach in the context of wrongs suffered by people at work is not normally associated with prosecution but is concerned with compensation. The large sums awarded by courts following successful claims resulting from injuries at work where negligence on the part of the employer can be shown (i.e. failure to carry out Common Law duty) underline the importance of compliance. There is anyway a moral duty to operate a place of work in such a way that accidents are avoided but the present concern is with the law.

There is an erroneous impression that compensation claims have little effect on an employer's economy on the grounds that insurance companies carry the burden. It is correct that legislation entitled the Employers' Liability (Compulsory Insurance) Act makes such insurance mandatory. It is untrue that expensive claims do not react on the insured employer. There is, of course, a degree of spreading the risk but almost all insurers apply a weighting factor to premiums levied. The fewer the claims the lower the premium and the converse is the case. This feature of employers' liability insurance is likely to receive even greater emphasis in the 1980s for a variety of reasons. The so-called Pearson Commission, set up to look at aspects of Common Law and make recommendations regarding changes in its application, produced in 1978 a report[12] which is mildly controversial and which will, if implemented, increase the number of compensation claims and enhance their chance of success. Much tighter product liability legislation which will provide a supportive role in compensation claims, will certainly have a similar effect although in a different but related field. An amendment to HASAWA contained in the Employment Protection Act[13] makes information in support of compensation claims more readily available, as do the Safety Representatives and Safety Committees Regulations.[14] The desirability of these changes or introductions is not questionable. The point to be made is that they may, and almost certainly will, place an added financial burden on employers. The remedies are to avoid the starting-point of such claims—the work accident, and to be aware of Common Law duty so that compliance can be strived for.

A concise summary of that duty was pronounced by Lord Oaksey[15] in the following terms: "The duty of an employer towards his servant is to take reasonable care for the servant's safety in all the circumstances of the case". This poses the question of what is meant by reasonable care. Reasonable practicability enters into consideration here as well as in statutory duty. If an employer exposes an employee to hazards which he can reasonably be expected to foresee and fails to take such steps as are reasonably practicable to protect the employee against such hazards, a court is likely to take the view that a failure to discharge Common Law duty has occurred. It may be that the employer is unaware of the existence of a hazard. The test then is, should he have been aware of it?

There does not have to be a breach of statutory duty for a claim for damages to succeed if it can be shown that the employer was negligent. An example in a laboratory in a chemical works is that while the Chemical Works Regulations 1922 require the provision of suitable hand and eye protection where strong acids are handled or used, there is no such requirement for body protection. But an employer is aware that there is a risk of injury to the body. Therefore a failure to protect this area by containment or personal protection would constitute a failure to comply with Common Law duty, even though no breach of the CWR occurred. It must be remembered, however, that a failure to protect in this way could be regarded as a breach of HASAWA 2(1).

Common Law duty can be compartmented in the following way. The employer must employ competent workers because the incompetent may negligently endanger their fellow employees. This duty is difficult to comply with because of the extreme care which must be used when dismissing even the incompetent employee after a period of service. The importance of careful initial selection is thus underlined, as is the rapid identification of substandard employees. Proper plant and equipment must be provided which is suitable for the purpose for which it is to be used. Improvisation should therefore be looked at very carefully. Flixborough is an example of improvisation which went wrong. Good and regular maintenance is important, too. The systems of work must be safe. It is not unusual to find unsatisfactory or outdated systems of work perpetuated because of the "We've done it this way safely for years" mentality. The system should be carefully thought out initially then examined from time to time to see if revision or updating is necessary. Finally, the place of work itself must be safe, as well as access to it and egress from it. The similarity between these and statutory duties will have been noted. Common Law duty is wider in the sense that it covers activities not mentioned in statutes, and is a matter between the parties involved.

REFERENCES

1. See "Guide to the HFL & LPG Regulations, HMSO Reg 3, lines 8 and 9.
2. Proceedings of the 1975 conference of the TUC.
3. Edwards vs. National Coal Board 1949.
4. See Pamphlet HSC 6 from the Health and Safety Executive.
5. Employment Protection (Consolidation) Act 1978.
6. Armour vs. Skeen: Scots Law Times 22.4.77.
7. Occupational Safety and Health, November 1979, pages 20 and 21.
8. The Fire Precautions (Non-Certificated Factory, Office, Shop and Railway Premises) Regulations 1976.
9. The Asbestos Research Council, Environmental Control Committee, 114 Park Street, London W1Y 4AB.
10. As in (1).
11. Health and Safety Booklet No. 13, HMSO, London.
12. The Royal Commission on Civil Liability and Compensation for Personal Injury (Pearson Commission).
13. Employment Protection Act 1975, Schedule 15 (9).
14. Safety Representatives and Safety Committees Regulations S.I. 1977, No. 500, Regulations 4 (1) (g) and 5 (3).
15. Paris vs. Stepney Borough Council, 1951.

FURTHER READING

"Questions and Answers on the Health and Safety at Work etc Act", Alan Osborne and Associates (Books) Ltd, Unit 5, Seager Buildings, Brookmill Road, London.

14

Safety Goals and How to Achieve Them

ESTABLISHING THE NEED FOR SAFETY GOALS

The concept of health and safety activities as something outside the orbit of laboratory workers other than those such as safety officers with a specific involvement, is slowly finding its way to its proper place, which is that particular mental wastebasket used for outdated and discarded ideas. Too many laboratory managers and section leaders had a very special niche into which they lodged their safety responsibilities. It was small in size and difficult to find so there was little obtrusion into the "important" elements of their managerial functions, such as seeing work programmes through, providing and stimulating a "think tank" for staff, and generally justifying their existence in a world which placed considerable emphasis on tangible results and not much on peripheral activities. Technicians and assistants came into laboratories ill prepared to work in circumstances and amongst substances which presented hazards of some magnitude. Anyone who contended that before the last two decades educational establishments gave high priority to safety training in laboratories viewed the situation through spectacles of a particularly rosy hue.

This situation has changed for the better at least in this respect for two major reasons. The first is that legislation has caught up with laboratories, the second that moral values have changed. It is an oddity that in a period when public manifestations of morality have taken the forms of riots, conflicts of other kinds, savagery in everyday life on a scale unthought of as recently as the 1950s and 1960s, there should at the same time be a quieter, more restrained and generally more private concern for individuals and their health and welfare. The "make advances and get the work done at any price" ideology has diminished and is disappearing. Genuine concern for the health and general well-being of employees is gradually superceding the unhappily still only partly obsolescent disregard for safety and the avoidance of accidents and injuries.

The facts and ideas (it is a mixture of both) expressed in this book can, if applied and developed, accelerate this progression from the frequently unwelcome working conditions of the first half of this century, not to perfection, but at least to a genuinely acceptable mode. But the greatest impetus towards the near-perfection which should be sought comes from the highest level of management. H. W. Heinrich, whose role in accident prevention can be likened to that of Sigmund Freud in psychoanalysis, illustrates this point in describing the high frequency and severity of employee accidents which existed in a group of chemical manufacturing plants and the accompanying laboratories.[1] When the problem became so acute that it merited consideration by top executives, these facts were identified.

Local managers were "too busy" to initiate and direct safety activities. Output was placed so far above safety in importance that the latter was practically ignored. A wrong conception existed concerning the relationship between safety and output and also as to the effect of accident cost on the particular enterprise's economy. At each location managers paid little attention to the advice of the safety officer. No attempt had ever been made to establish orderly and effective safety programmes, nor was this likely to occur unless some drastic action was taken at a level which would be heeded. Eventually the president of the parent company issued a direct personal order in the following terms:

1. The senior executive to establish immediately a central safety committee.
2. The senior executive himself to act as chairman of the committee.
3. In each location a safety committee to be formed and a safety inspector appointed.
4. On every site the senior manager to spend one full day each week wholly on accident prevention work.
5. That necessary forms be provided, accidents properly investigated, job analyses to be made, responsibility to be placed, safeguards installed, and in general that safety work be carried on in the future in an orderly, systematic, and effective manner.

The president also laid down that the senior manager on each site was to make a personal report at stated intervals to the head office of the parent company. Not surprisingly, managers were no longer "too busy" to attend to their safety responsibilities thereafter. Astonishing improvements in safety performance followed, accompanied by rises in the quality and volume of output—there is something of the sycophant in almost everyone. The experience recounted above is not an isolated event.

IMPETUS FROM CHIEF EXECUTIVES

The authors have worked for many years for an international company which had a succession of chief executives, each of whom was determined that the enterprise would be conducted with the maximum attention given to the health and safety of employees. Besides control laboratories on each plant, the technical headquarters included a laboratory complex covering $2\frac{1}{2}$ acres where research and development work was carried on as well as a technical service for customers. Year after year the company achieved outstanding safety performance, consistently heading the safety league table compiled by trade associations in the countries in which operations were located. During the same period the company excelled in regard to every index of efficiency—profitability, production, and quality. This underlines yet again an essential truth; where the chief executive is determined that high levels of health and safety will be achieved and maintained, this attitude will permeate the whole structure of the enterprise. Where there is a lack of determination and interest at the top strata of management the reverse is the case. There is nothing novel or innovative about this. It is equivalent to saying that if tidal estuaries are continuously dredged the navigation channels will not silt up. The difference is that harbour authorities accept that dredging is an essential operation if their business is to operate successfully, whereas some senior managers in other spheres do not appreciate that failure to give a lead to health and safety activities can have an injurious effect on the enterprise they control.

AN ACTION PLAN

Carrying out a Safety Audit

Assuming that sub-standard conditions exist or improvements are sought, the procedure should be to initiate a safety audit. This term is used to describe a range of inspection techniques, linked to a programme of remedial action formulated as a result of what the inspection reveals. The term is borrowed from the field of accountancy and is particularly apt if one takes part of the Oxford English Dictionary's definition of audit—"a searching examination". A safety audit considers the total laboratory environment.[2] Each area is subjected to a critical examination with the object of minimizing injury, damage,or loss. Every component of the total system is included under the following headings:

- Management policy
- Attitudes
- Training
- Features of procedure, including design, layout, and construction
- Emergency plans
- Personal protection standards
- Accident investigation
- Accident records.

Information covering each of these headings is contained in this book. The audit identifies strengths and weaknesses and the main areas of vulnerability. It should not be thought that the technique can be applied only to multi-section laboratories or large complexes. It can be scaled down to apply even to the single laboratory. One of its merits is that it does not rely on failures as the basis for measuring the laboratory safety effort, as does accident investigation, for example.

Setting up the Audit Team

The first step is for the laboratory manager/chief chemist to decide the make-up of the audit team. Included may be the safety officer if there is such an appointee, a senior chemist, a laboratory assistant, and a member of the safety committee. Before the audit begins there should be a thorough briefing so that members of the team unfamiliar with surveillance techniques understand the procedure and objectives. Where someone knowledgeable on the subject is not available the information in this chapter can be used. The following should be explained:

a) the overall objectives;
b) methods of data collection and recording;
c) the need for, and compilation of, a check list;
d) how to analyse the results;
e) the Action Plan which emerges from the audit.

Identifying Objectives

The overall objectives can be described as agreeing on standards of health and safety that are the eventual goals; identifying where there are shortfalls; and devising an action plan which, when implemented, will enable the agreed goals to be achieved.

Data collection and recording should be by careful personal observation and questioning, the inspection of records, and discussion with laboratory management, supervision, and representatives of lower echelons of employees. A carefully compiled check-list setting down the items to be examined will transform a somewhat aimless wandering around into a meaningful and purposeful examination. Items mentioned will include condition of premises and equipment, training, fire prevention and protection, procedures, accident investigation and records, storage, handling of substances, involvement of employees in the safety programme, protective clothing and devices, housekeeping, security, and assessment of current, potential and long-term risks. Other headings may be needed related to the particular tasks carried out. The further division of item headings is fairly obvious, but here is an example:

Premises and equipment

Suitability. Structural condition. Age. Workshops. Offices. Laboratories. Stores. Ancillary buildings (canteens, toilets, etc.). Maintenance standards. Neighbouring premises (nature, proximity, possibility of fire spread, possible effect of chemical leaks).

There should be space under each heading for additional items to be added as revealed during the audit.

Analysing Results

The analysis of the results should be conducted by the audit team in company with section managers and supervisors who would be involved only in so far as their own area of control is concerned. Their temporary presence is not to divert or influence the audit team's findings but to answer questions and provide additional information as the need arises. The completed check-lists will be compared and an action plan prepared.

Allocating Priorities

This should allocate priorities as follows, related to the urgency of action needed:

a) immediate;
b) as soon as possible;
c) when practicable;
d) when convenient.

A low priority should not be an excuse for inaction. The allocation is intended to ensure that resources are first used where there is the greatest need. It is not always possible to deal immediately with a) priorities, in which case temporary action can reduce them to a lower category. For example, it may be decided that in an area where normal electrical circuits and equipment are installed work is carried out involving flammable or potentially explosive substances so suitable flameproof equipment should be installed as a high priority. The procurement of this equipment and the changeover will, it is found, take some weeks or even months. The work concerned can temporarily be moved to a safe area, undertaken in hired premises, carried out under contract by others, or suspended until the original area can be made safe. Whatever course of action is chosen, the vital element is that in as short a time span as possible the action plan is implemented. An example of priority c) could be where lighting improvements are agreed. The laboratory concerned may be due for redecoration in the near future, in which case it is sensible to improve the lighting at the same time, thus disrupting work as little as possible provided that the delay does not subject employees to hazardous conditions in the meantime.

A common fault noted in audit teams is a tendency to concentrate on physical as opposed to people aspects—the hardware instead of the software of audits to use the jargon which has inevitably accompanied the development of the technique. Careful questioning will reveal the kind of attitude which prevails, the level of involvement in safety programmes, whether training is effective or otherwise, and if there is a genuine desire on the part of managers and supervisors to ensure that health and safety activities are treated with the importance they merit.

Future Exercises

Following an initial audit the frequency of subsequent exercises should be agreed in the light of experience gained. A full audit may not be necessary on future occasions. Safety surveys[3] (detailed thorough examinations of a specific field of activity), safety inspections (routine inspections of laboratories carried out by staff), safety tours (unscheduled examinations of the work area carried out by a laboratory or section manager), and hazard and operability studies (a formal critical examination of laboratory situations, especially where new layout or change of function is concerned) are activities which will supplement the audit. It may sometimes be desirable to call in an outside specialist where unfamiliar substances are used, or where projects are envisaged.

All the above suggestions are of a quasi-formal nature. Their adoption cannot fail to raise health and safety standards provided that the defects identified are remedied. If they are not, the whole safety organization falls into disrepute.

Finally, an incomplete understanding of the meaning of the words "hazard" and "risk" is frequently encountered. Often one of these words is used when what is really meant is the other. Hazards are inherent in every aspect of life. At home one may fall down the stairs, or slip on polished floors. On the road a car may skid because of slippery matter deposited on the driving surface. In laboratories hazards are present in the physical conditions, the substances handled, the systems or work, and so on.

Risks are the possibility of these hazards giving rise to accidents or injuries. Even though so long as human intervention is possible their complete elimination is unattainable, risks can be managed and controlled. Hopefully this book will help those who work in chemical laboratories to achieve these objectives.

REFERENCES

1. Heinrich, H. W. (1950). "Industrial Accident Prevention—a Scientific Approach", McGraw-Hill, New York and London. Heinrich was an eminently practical man; it was said when he was posthumously elected to the National Insurance Hall of Fame in 1980 "his life's work helped lay the foundation for the modern practice of safety engineering and accident prevention . . ."
2. "Laboratory Safety Management", Chemical and Allied Products Industry Training Board, Staines, England.
3. "Safety Audits", Chemical Industries Association, London, England.

Additional Bibliography

Alderson, R. H. (1975). "Design of the Electron Microscope Laboratory", North-Holland Publishing Co., Amsterdam/Oxford. American Elsevier Publishing Co., New York.

American Conference of Governmental Hygienists, Documentation of Threshold Limit Values for Substances in the Work Room Air, 1980.

Applying TLV's (1980). *Occupational Health* **32,** No. 6, 301.

Berman, E. (1980). "Toxic Metals and their Analyses", Heyden, London.

Boursnell, J. C. (1959). "Safety Techniques for Radioactive Tracers", Cambridge University Press.

Brenner, M. (1976). "Hazards of School Science", *New Scientist* **69,** 550–552.

Bretherick, L. (1975). "Handbook of Reactive Chemical Hazards", Butterworths, London.

Browning, E. (1965). "Toxicity and Metabolism of Industrial Solvents", Elsevier Publishing Co., New York.

Browning, E. (1953). "Toxic Solvents", Edward Arnold and Co., London.

Browning, E. (1961). "Toxicity of Industrial Metals", Butterworths, London.

Chemical Safety Data Sheets, Manufacturing Chemists Association, Washington DC, USA.

Clark, R. P. (1980). The evaluation of open-fronted biological "safety" cabinets, *Laboratory Practice* **29,** No. 9, 926.

"Code of Practice for Chemical Laboratories" (1976). Royal Institute of Chemistry, London.

"Code of Practice Against Radiation Hazards", 6th Edition (1973). Imperial College, London.

Cooke, J. (1979). "Perchloric Acid, the Dangers of Contamination in the Laboratory", *Health & Safety at Work* **54.**

Elkins, H. B., 2nd Edition (1959). "The Chemistry of Industrial Toxicology", John Wiley & Sons, London and New York.

Ellis, J. G. and Riches, N. J. (1978). "Safety and Laboratory Practice", Macmillan Press, London.

"An Encyclopaedia of Chemicals and Drugs", 9th Edition, The Merck Index, Merck & Co.

Everett, K. (1966). "University of Leeds Safety Handbook", University of Leeds Safety Committee.

Fairchild, E. J. (1978). "Suspected Carcinogens", Castle House Publications Ltd, London.

Fawcett, H. H. and Wood, W. S., eds (1965). "Safety and Accident Prevention in Chemical Operations", Interscience Publishers, London.

Ferguson, W. R. (1973). "Practical Laboratory Planning", Applied Science Publishers Ltd, London.

"Fisher Manual of Laboratory Safety" (1972). Fisher Scientific Co., Pittsburgh, USA.

Gemert, L. J. van and Nettenbreijer, A. H. (1977). "Compilation of Odour Thresholds in Air and Water", National Institute for Water Supply, Voorburg; Central Institute for Nutrition and Food Research, TNO Zeist, The Netherlands.

Gosselin, R. *et al.* (1976). 4th Edition, "Chemical Toxicology of Commercial Products", Williams and Wilkins Co.

Goyer, R. A. and Melhman, M. A. (1977). "Toxicology of Trace Elements", John Wiley & Sons, London.

Green, M. E. and Turk, A. (1978). "Safety in Working with Chemicals", Macmillan Publishing Co., New York.

Grover, F. and Wallace, P. (1979). "Laboratory Organisation and Management", Butterworths, London.

"Guidance Notes for the Protection of Persons Exposed to Ionising Radiations in Research and Teaching", 2nd Edition (1968), HMSO, London.

"Guide for Safety in the Chemical Laboratory", 2nd Edition (1972). Van Nostrand Reinhold Co., New York.

"A Guide to the Safe Use of X-ray Crystallographic and Spectrometric Equipment" (1977). Association of University Radiation Protection Officers, Monograph Series No. 1, P. J. F. Griffiths, University of Wales Institute of Science and Technology, Cardiff.

Guy, K. (1962). "Laboratory Organisation and Administration", MacMillan Press, London.

Hackett, W. J. and Robbins, G. P. (1979). "Safety Science for Technicians", Longman, London and New York.

"Handling Chemicals Safely", 2nd Edition (1980). Dutch Association of Safety Experts, Dutch Chemical Industry Association, Dutch Safety Institute.

Hawley, G. A. (1977). "Condensed Chemical Dictionary", Van Nostrand Reinhold Co., New York.

"Health and Safety in Welding", 2nd Edition (1965). The Institute of Welding, London.

Hughes, D. (1974). The design and installation of efficient fume cupboards, *British Journal of Radiology* **47,** 888–892.

Hughes, D. (1980). "Literature Survey and Design Study of Fume Cupboards and Fume Disposal Systems", Science Reviews Ltd, London.

"Identification Colours for Pipes Conveying Fluids in Liquid or Gaseous Condition in Land Installations and on Board Ships" (1966). ISO Recommendation R 508, International Organization for Standardization.

"Ionizing Radiations, Supplementary Proposals for Provisions on Radiological Protection and Draft Advice from the National Radiological Protection Board to the Health and Safety Commission" (1979). HMSO, London.

"Ionizing Radiations, Proposals on Radiological Protection" (1979). HMSO, London.
"The Ionizing Radiations (Unsealed Radioactive Substances) Regulations, 1968. The Ionizing Radiations (Sealed Sources) Regulations, 1969, Radioactive Substances Act, 1960. HMSO, London.
"Laboratory Safety Handbook" (1976). Sanderson Chemical Consultants Ltd, Cleveland, UK.
"Laboratory Safety Management" (1979). Chemical and Allied Products Industry Training Board, England.
"Living with Radiation" (1973). National Radiological Protection Board, HMSO, London.
Mann, C. A. (1969). "Safety in the Chemical Laboratory, LVIII, Science Experiment Safety in the Elementary School", *Journal of Chemical Education* **46,** Part 5, A347–A353.
Martindale, "The Extra Pharmacopeia", 27th Edition (1977). Pharmaceutical Press, London.
Mechanical Safety Devices for Laboratory Centrifuges, BS 4402: 1969.
"Model Code of Principles of Construction and Licensing Conditions (Part II) for Distributing Depots and Major Installations" (1968). HMSO, London.
Paget, G. E., ed. (1979). "Topics in Toxicology—Good Laboratory Practice", MTP Press Ltd, London.
"Phosgene" (1975). Codes of Practice for Chemicals with Major Hazards, The Chemical Industry Safety and Health Council, Chemical Industries Association, London.
Pieters, H. A. J. and Creyghton, J. W. (1975). "Safety in the Chemical Laboratory", 2nd Edition, Butterworths, London.
Plunkett, E. R. (1976). "Handbook of Industrial Toxicology", Heyden, London.
The Poisons Rules, SI, HMSO, London.
"Proceedings of the International Symposium on Maximum Allowable Concentrations of Toxic Substances in Industry" (1961). Butterworths, London.
"Radiological Protection in the Universities" (1966). The Association of Commonwealth Universities, London.
"Recommendations of the International Commission on Radiological Protection" (1965). Pergamon Press, Oxford, UK.
"Recommendations on Laboratory Furnishings and Fittings", BS 3202: 1959 (under revision), British Standards Institution, London.
"Recommended Safety Precautions for Handling Cryogenic Liquids" (1975). BOC Ltd, Edwards High Vacuum, Crawley, UK.
"Registry of Toxic Effects of Chemical Substances" (1978). National Institute for Occupational Safety and Health, US Department of Health, Education and Welfare, Washington DC, USA.
"Safety in the Chemical Laboratory" (1966). May and Baker Ltd, Dagenham, UK.
"Safety in Chemical Laboratories and in the use of Chemicals" (1970). Imperial College of Science and Technology, London.
"Safety in Science Laboratories, 3rd Edition (1978). HMSO, London.
Sax, N. I. (1979). "Dangerous Properties of Industrial Materials", 5th Edition,

Van Nostrand Rheinhold Co., New York.

Schuerch, C. (1972). Safe practice in the chemical laboratory, **49,** Ay683–685, 637–A639.

Scott, R. B. and Hazar, A. S. (1978). Liability in the academic chemical laboratory, *Journal of Chemical Education*, **55,** Part 4, A196–A198.

"Specification for Identification of Pipelines" (1975). BS 1710, British Standards Institution, London.

Stalzer, R. F., Martin, J. R. and Railing, W. E. (1967). Safety in the laboratory. *In* "Treatise on Analytical Chemistry", Part III, Vol. 1, Interscience Publishers, London.

"Static Electricity" (1963). Safety in Industry, Mechanical and Physical Hazards No. 8, Bulletin No. 256, US Department of Labor, Bureau of Labor Standards, Washington DC, USA.

Steere, N. V. (1971). "Handbook of Laboratory Safety", C.R.C. Press Inc., Florida, USA.

Taylor, P. J. (1978). "Developing an OSHA Acceptable Academic Chemistry Department", *Journal of Chemical Education*, **55,** Part 12, A439–A441.

"Threshold Limit Values for 1980", Guidance Note EH 15/70, HMSO, London.

"Toxic and Hazardous Industrial Chemical Safety Manual" (1976). The Information Institute, Tokyo.

Toxic Metals and their Analysis (1980). Berman, E.,Heyden, London.

Wawzonek, S. (1978). Safety practice in the undergraduate organic chemistry laboratory, *Journal of Chemical Education*, **55,** Part 2, A71–A74.

Weart, R. C., ed., 61st Edition (1980). "Handbook of Chemistry and Physics", The Chemical Rubber Publishing Co., Cleveland, USA.

Young, J. R. (1971). Responsibility for a safe high school laboratory, *Journal of Chemical Education*, **48,** A349–A356, and Survey of safety in high school chemical laboratories, **47,** Part 12, A829–A838.

Zabetakis, M. G. (1967). "Safety with Cyogenic Fluids", Haywood Books, London.

Useful Addresses

A. Gallenkamp & Co. Ltd.,
P.O. Box 290,
Technico House,
Christopher Street,
London EC2P 2ER.

Albert E. Marston & Co. Ltd.,
Wellington Works,
Planetary Road,
Willinghalls,
West Midlands WV13 3ST.

American Conference of Governmental Industrial Hygienists Inc.,
6500 Glenway Avenue,
Building D-5,
Cincinnati,
Ohio 45211, USA.

British Standards Institution,
2 Park Street,
London W1A 2BS.

CRC Press Inc.,
2000 N.W. 24th Street,
Boca Raton,
Florida 33421, USA.

DIN Deutsches Institut für Normung e.V.,
Burggrafenstrasse 4-10,
Postfach 1107.
D-1000 Berlin 30,
Germany.

Dow Corning Ltd.,
Reading Bridge House,
Reading,
Berks. RG1 8PW.

Efficiency Aids Ltd.,
Prospect House,
Norton Road,
Stockton-on-Tees,
Cleveland.

Health & Safety Executive,
Baynards House,
1 Chepstow Place,
London W2 4TF.

Imperial Chemical Industries Ltd.,
Imperial Chemical House,
Millbank,
London SW1P 3JF.

Mine Safety Appliances Co. Ltd.,
East Shawhead,
Coatbridge ML5 4TD.

Mini-instruments Ltd.,
8 Station Industrial Estate,
Burnham-on-Crouch,
Essex CMO 8RN.

The Narda Microwave Corporation,
Plainview L.I.,
New York 11803, USA.

National Radiological Protection Board,
Harwell, Didcot,
Oxon. OX11 0RQ.

Philips Industries,
Scientific and Industrial Equipment Division,
5600 MD Eindhoven,
The Netherlands.

The Royal Society of Chemistry,
Burlington House,
London W1V 0BN.

Walter Page (Safeways) Ltd.,
46 Lower Shelton Road,
Marston Moretaine,
Bedford, MK43 0LW.

Index